ONLY ONE OCEAN

Teacher's Guide

Grades 5–8

Skills

Recording • Drawing Conclusions • Working Cooperatively • Graphing
Observing • Inferring • Making Models • Researching • Comparing
Dissecting • Extracting Meaning from Text • Communicating
Logical and Critical Thinking

Concepts

Interconnectedness of Ocean Basins • Respect for Organisms • Oceanography
Physical Geography • Upwelling Zones • Ocean Resources • Adaptation
Form and Function • Overfishing • Alternative Perspectives
Consumer and Environmental Awareness

Themes

Models and Simulations • Systems and Interactions • Patterns of Change
Interdependence • Abundance and Diversity • Scale and Structure

Mathematics Strands

Number • Measurement • Pattern • Logic and Language • Data Interpretation
Ratio and Proportion • Drawing to Scale • Rational Numbers
Symmetry • Statistics and Probability

Nature of Science and Mathematics

Creativity and Constraints • Interdisciplinary Connections
Real-Life Applications

by
Catherine Halversen and **Craig Strang**
with **Kimi Hosoume**

Great Explorations in Math and Science
Lawrence Hall of Science
University of California at Berkeley

Cover Design
Lisa Haderlie Baker

Design and Illustrations
Lisa Klofkorn

Photographs
Carl Babcock
Craig Strang

Lawrence Hall of Science, University of California,
Berkeley, CA 94720-5200

Director: Ian Carmichael

Publication of *Only One Ocean* was made possible by a grant from the Employees Community Fund of Boeing California and the Boeing Corporation (originally the McDonnell-Douglas Foundation and Employee's Community Fund). The GEMS Program and the Lawrence Hall of Science greatly appreciate this support.

Initial support for the origination and publication of the GEMS series was provided by the A.W. Mellon Foundation and the Carnegie Corporation of New York. Under a grant from the National Science Foundation, GEMS Leader's Workshops have been held across the country. GEMS has also received support from the McDonnell-Douglas Foundation and the McDonnell-Douglas Employee's Community Fund; Employees Community Fund of Boeing California and the Boeing Corporation; the Hewlett Packard Company; the people at Chevron USA; the William K. Holt Foundation; Join Hands, the Health and Safety Educational Alliance; the Microscopy Society of America (MSA); the Shell Oil Company Foundation; and the Crail-Johnson Foundation. GEMS also gratefully acknowledges the contribution of word processing equipment from Apple Computer, Inc. This support does not imply responsibility for statements or views expressed in publications of the GEMS program. For further information on GEMS leadership opportunities, or to receive a catalog and the *GEMS Network News*, please contact GEMS at the address and phone number below. We also welcome letters to the *GEMS Network News*.

Printed on recycled paper with soy-based inks.

International Standard Book Number: 0-924886-22-6

COMMENTS WELCOME !

Great Explorations in Math and Science (GEMS) is an ongoing curriculum development program. GEMS guides are periodically revised to incorporate teacher comments and new approaches. We welcome your suggestions, criticisms, and helpful hints, and any anecdotes about your experience presenting GEMS activities. Your suggestions will be reviewed each time a GEMS guide is revised. Please send your comments to: GEMS Revisions, c/o Lawrence Hall of Science, University of California, Berkeley, CA 94720-5200. The phone number is (510) 642-7771 and the fax number is (510) 643-0309. You can also reach us by e-mail at gems@uclink.berkeley.edu, or visit our Web site at www.lhsgems.org.

Great Explorations in Math and Science (GEMS) Program

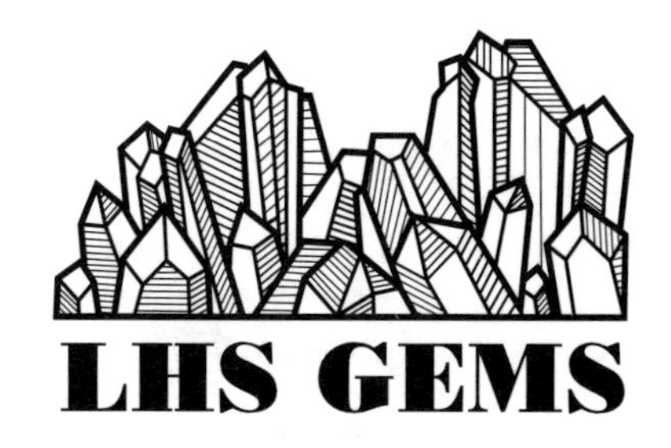

The Lawrence Hall of Science (LHS) is a public science center on the University of California at Berkeley campus. LHS offers a full program of activities for the public, including workshops and classes, exhibits, films, lectures, and special events. LHS is also a center for teacher education and curriculum research and development.

Over the years, LHS staff has developed a multitude of activities, assembly programs, classes, and interactive exhibits. These programs have proven immensely successful at the Hall and should be useful to schools, other science centers, museums, and community groups. A number of these guided-discovery activities have been published under the Great Explorations in Math and Science (GEMS) title, after an extensive refinement and adaptation process that includes classroom testing of trial versions and modifications to ensure the use of easy-to-obtain materials. Carefully written and edited step-by-step instructions and background information allow presentation by teachers without special background in mathematics or science.

Staff

Director: Jacqueline Barber
Associate Director: Kimi Hosoume
Associate Director/Principal Editor: Lincoln Bergman
Mathematics Curriculum Specialist: Jaine Kopp
GEMS Network Director: Carolyn Willard
GEMS Workshop Coordinator: Laura Tucker
Staff Development Specialists: Lynn Barakos, Katharine Barrett, Kevin Beals, Ellen Blinderman, Gigi Dornfest, John Erickson, Stan Fukunaga, Debra Harper, Linda Lipner, Karen Ostlund
Distribution Coordinator: Karen Milligan
Workshop Administrator: Terry Cort
Trial Test and Materials Manager: Cheryl Webb
Financial Assistant: Vivian Tong
Distribution Representative: Felicia Roston
Shipping Assistant: Maureen Johnson
Director of Marketing and Promotion: Matthew Osborn
Senior Writer: Nicole Parizeau
Senior Editor: Carl Babcock
Editor: Florence Stone
Principal Publications Coordinator: Kay Fairwell
Art Director: Lisa Haderlie Baker
Senior Artists: Carol Bevilacqua, Rose Craig, Lisa Klofkorn
Staff Assistants: Haleah Hoshino, Mikalyn Roberts, Thania Sanchez, Stacey Touson

Contributing Authors

Jacqueline Barber
Katharine Barrett
Kevin Beals
Lincoln Bergman
Susan Brady
Beverly Braxton
Kevin Cuff
Linda De Lucchi
Gigi Dornfest
Jean C. Echols
John Erickson
Philip Gonsalves
Jan M. Goodman
Alan Gould
Catherine Halversen
Debra Harper
Kimi Hosoume
Susan Jagoda
Jaine Kopp
Linda Lipner
Larry Malone
Cary I. Sneider
Craig Strang
Herbert Thier
Jennifer Meux White
Carolyn Willard

Reviewers

We'd like to thank the following educators, who reviewed, tested, or coordinated the reviewing of this series of GEMS guides (*Only One Ocean* and *Ocean Currents*—originally bound together as *Open Ocean*—and *Environmental Detectives)* in manuscript or draft form. Their critical comments and recommendations, based on classroom presentation of these activities nationwide, contributed significantly to these GEMS publications. Their participation in the review process does not necessarily imply endorsement of the GEMS program or responsibility for statements or views expressed. Their role is an invaluable one; feedback is carefully recorded and integrated as appropriate into the publications. THANK YOU!

ARIZONA

Sonoran Sky Elementary School, Scottsdale
Brent Engilman
Marge Maceno *
Amy Smith
Kathy Wieeke
Tammy Wopnford

ARKANSAS

Bob Courtway Middle School, Conway
Robin Cole
Rick Hawkins
Charlcie Strange *
Paula Wilson

Birch Kirksey Middle School, Rogers
Beth Ann Carnes
Jenny Jones *
Curtis S. Smith
Sharron Wolf

CALIFORNIA

Albany Middle School, Albany
Karen Adams
Jenny Anderson
Cyndy Plambeck
Kay Sorg

Martin Luther King Middle School, Berkeley
Indigo Babtiste
Akemi Hamai
Yvette McCullough
Beth Sonnenberg

Harding School, El Cerrito
Renie Gannett
Carol Leitch
Jim Wright

Portola Middle School, El Cerrito
Debbie Marasaki
Carol Mitchell
Susan Peterson
Mike Wilson

Warwick Elementary School, Fremont
Dale Harden
Katy Johnson
Richard Nancee
Robert Nishiyam
Bonnie Quigley
Ann Trammel

Ohlone Elementary School, Hercules
Stacey Cragholm
Gloria Crim
Jay Glesener
Sandra Simmons

Hall Middle School, Larkspur
Trish Mihalek
Art Nelson
Ted Stoeckley
Barry Sullivan

McAuliffe Middle School, Los Alamitos
Michelle Armstrong
Kathy Burtner

Oak Middle School, Los Alamitos
Joyce Buehler
Rob Main *
Kendall Vaught

Hidden Valley School, Martinez
Diane Coventry
Nigel Dabby
Jennifer Sullivan

Calvin Simmons Jr. High School, Oakland
Stan Lake
Wendy Lewis
Fernando Mendez
Thelma Rodriguez

Vintage Parkway School, Oakley
Jennifer Asmussen
Alisa Haley
Casey Maupin
Lian McCain
Steve Williams

Collins Elementary School, Pinole
Ralph Baum
Craig Payne
Anne Taylor
Genevieve Webb

Adams Middle School, Richmond
Richard Avalos
Susan Berry
John Eby
John Iwawaki
Steve Stewart

Bell Jr. High School, San Diego
Nick Kardouche
Elouise King *
Denise Vizcarra
Mala Wingerd

Dingeman Elementary School, San Diego
Monka Ely *
Godwin Higa
Kim Holzman
Linda Koravos

Cook Middle School, Santa Rosa
Steve Williams

Rincon Valley Middle School, Santa Rosa
Sue Lunsford
Penny Sirota *
Laurel VarnBuhler

FLORIDA

Howard Middle School, Orlando
Elizabeth Black
Carletta Davis
Susan Leeds *
Jennifer Miller

MISSOURI

St. Bernadette School, Kansas City
Brett Coffman
Dorothy McClung
Aggie Rieger
Margie St. Germain

Poplar Bluff 5th/6th Grade Center, Poplar Bluff
Cindy Gaebler
Leslie Kidwell
Barbara King *
Melodie Summers

NEVADA

Churchill County Jr. High School, Fallon
Kerri Angel
Deana Madrasco
Amy Piazzola
Sue Smith-Ansotegui *

NEW HAMPSHIRE

Crescent Lake Elementary School, Wolfeboro
Kate Borelli
Amy Kathan
Elaine M. Meyers *
Patti Morissey

NEW JERSEY

Orchard Hill School, Skillman
Jay Glassman *
Al Hadinger
Georgiana Kichura
Tony Tedesco

NEW YORK

Maple Hill Middle School, Castleton
Beth Chittendo
Jeanne Monteau *

Lewisboro Elementary School, South Salem
Debra Jeffers

St. Brigid's Regional Catholic School, Wateruliet
Patricia Moyles

OHIO

Baker Middle School, Marion
Dave Dotson
Denise Z. Iams *
Betty Oyster
Carol White

OREGON

Sitton School, Portland
David Lifton
Deborah Nass

TEXAS

Colleyville ISD, Grapevine
Kathy Keeney

Grapevine ISD/ Administration Building, Grapevine
Shelly Castleberry
Terry Dixon
Malanie Gable *
Randy Stuempfig

Ector County ISD, Odessa
Becky Stanford

Ireland Elementary School, Odessa
Susan Hardy

Miliam Elementary School, Odessa
Eli Tavarez

Travis Elementary School, Odessa
Patty Calk
Stacey Hawkins

WASHINGTON

The Gardner School, Vancouver
Matt Karlsen
Tom Schlotfeldt
Rob VanNood

* Trial test coordinators

Acknowledgments

We wish to thank the following people for their generosity and willingness to share their expertise and time to ensure that *Only One Ocean* is scientifically accurate and conceptually meaningful in the "real world."

Tom Murphree, Ph.D., Naval Postgraduate School, Monterey, California provided the scientific review of Activity 1, "Apples and Oceans," and made many important edits and comments to early versions. Throughout the process, Tom helped us refine our understanding of "what makes the ocean go"—which in turn helped describe the concepts more clearly to teachers and students alike.

The initial inspiration for the activity "Apples and Oceans" came from Dr. Stephanie Kaza of the University of Vermont, who helped write the very first version (oh so long ago). Francisca Sanchez's expertise in language acquisition helped hone "Apples and Oceans" into one of the foremost Specially Designed Academic Instruction in English (SDAIE) activities in the country.

Dr. Clyde Roper of the Smithsonian Institution provided the scientific review for Activity 2 and other squid portions of *Only One Ocean*. Dr. Roper graciously read the manuscript, provided many important edits and fascinating anecdotes, *and* wrote the foreword to this guide—all on the cusp of his departure for Australia in search of the giant squid. A tremendous thank you to Dr. Roper for his grace under pressure and his enthusiasm for sharing his expertise with the authors and editor of this guide (all of whom are now squid groupies).

Dr. Roger Hanlon, from Woods Hole Marine Laboratory, provided input on the squid illustrations. His book, *Cephalopod Behaviour* (Roger Hanlon and John Messenger), was an invaluable resource in writing this activity; the authors of *Only One Ocean* have waited a long time for so definitive a resource on these incredible organisms.

The fisheries activity, "What's the Catch?," was expertly reviewed by Michael McGowan, Ph.D., Fishery Oceanographer and Senior Research Scientist/Adjunct Professor of Biology, San Francisco State University, California. Dr. McGowan also provided thought-provoking insights on the issues of overfishing and sustainability, and wisely pointed out: "Natural resource management is a difficult problem with serious conflicts of interest all around;

otherwise it would have been solved long ago." Thanks to Dr. McGowan, more students will have the knowledge and willingness to tackle the problems of sustainable fishing, armed with new insights and ideas.

Scientists from the National Marine Fisheries Service (NMFS), under the auspices of the National Oceanic and Atmospheric Administration (NOAA), provided invaluable assistance and expertise regarding the walleye pollock fishery in Alaska. We would especially like to thank James Ianelli, Fisheries Resource Scientist; Andrew Smoker, Senior In-Season Manager, Alaska Region; and Jeff Passer, Enforcement Division Special Agent in Charge, Alaska Region, for their gracious responses to questions about pollock—"the most-fished species in the world." We're very grateful to Russell Oates, Chief of the Waterfowl Management Branch, U.S. Fish and Wildlife Service, for generously and diligently pointing us to the best possible sources.

The authors also feel very fortunate that Dr. E.C. Haderlie of the Naval Postgraduate School, Monterey, California read the entire guide for scientific accuracy.

A very special thank you goes to Nicole Parizeau, GEMS Senior Writer. Nicole's tireless and creative efforts and her insightful editing helped make this guide a reality and the best it could possibly be. We're also indebted to Jaine Kopp, GEMS Mathematics Curriculum Specialist, who plunged devotedly into the math and graphing aspects of the guide to ensure their accuracy and relevance. Lincoln Bergman, GEMS Principal Editor and Associate Director, participated in creating the homework assignment for Activity 3, favored us with his "Fishery Rhymes," and lent his reliable wisdom to the overall process. GEMS Curriculum Specialist Kevin Beals made inventive contributions to Activity 3, and Editor Florence Stone provided valuable organization to complex sections of the guide.

I do not know what I may appear to the world; but to myself I seem to have been only like a boy playing on the seashore, and diverting myself in now and then finding a smoother pebble or a prettier shell than ordinary, whilst the great ocean of truth lay all undiscovered before me.

Isaac Newton
(1642–1727)

Contents

Only One Ocean was developed by the Marine Activities, Resources & Education (MARE) program of the Lawrence Hall of Science. MARE (pronounced like the Latin word for ocean: mär´a) is an exciting, whole-school, interdisciplinary, marine science program for elementary and middle schools. This GEMS Teacher's Guide is excerpted and adapted from the *MARE Teacher's Guide to the Open Ocean*.

Please see page 8 for more information on MARE programs. In adapting portions of the MARE activities, this guide features a number of teaching strategies and "activity structures" developed by MARE and embedded in the MARE approach. These special methods are designed to encourage the acquisition of both science and language, especially among English-language learners. They also foster cooperative learning. They're discussed in more detail in the main introduction of this guide, and are described step by step as part of the activities.

Foreword

The Ocean. That vast, complex, watery engine that drives Planet Earth. It is responsible for global climate, creates continental as well as oceanic weather, and ultimately controls all biological productivity on Earth. Without the ocean—our "Inner Space"—the water planet would cease to exist. All life would vanish.

While we know something about the edges of the ocean, where shallow water meets the land, and about the waters along the continental shelf, we know very little detail of the ocean's deeper waters. In fact, some marine scientists are quite correct to insist that we know more about Outer Space than we do about Inner Space, our own ocean. The deep ocean is so little explored, so vast, so inaccessible, so complex, so hidden from our sight and mind, that we must now dedicate our attention and research to the depths, to gain knowledge that will ensure our long-term survival.

The best place to begin to inform people about the ocean is in school, where eager students can become the ocean-minded generation. But the size and complexity of the ocean are so great that we need focal points through which to introduce students to the sea, to spark their attention and interest and to secure their dedication to learning. Using the squid as an icon to represent the oceans, as Activity 2 of this guide does, can be one of those focal points. Squids are fascinating animals. They occur in all ocean basins of the world and make superb ambassadors to the sea, representing and introducing the ocean's neighbors and neighborhoods.

Only One Ocean, in addressing the interconnectedness of our ocean, accomplishes the difficult goal of covering a very broad and complex topic in an accessible way.

Clyde F.E. Roper
Department of Invertebrate Zoology
National Museum of Natural History
Smithsonian Institution

April 23, 2001

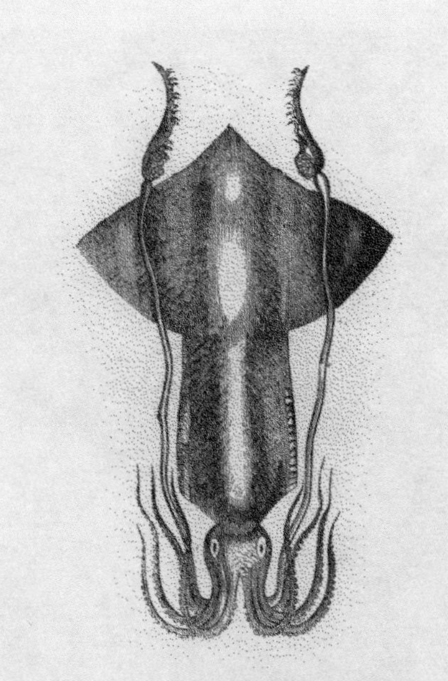

"That's too bad," said Conseil. "I'd like to come face to face with one of these giant squid I've heard so much about, and which can drag ships down to the bottom. They're called kraken, I think."

"You'll never get me to believe such animals exist," said Ned Land....

"Yes, but others still believe in it."

"Maybe they do, Conseil, but as for me, I've decided not to admit these monsters really exist until I've dissected one with my own hands."

"Doesn't Monsieur believe in giant squid either?" Conseil asked me.

"Who in the name of heaven has believed in them?" cried the Canadian.

"Lots of people, Ned, my friend."

"Maybe scientists, but not fishermen."

"No, Ned. Both fishermen and scientists!"

—from *20,000 Leagues Under the Sea* by Jules Verne.

(See "Literature Connections" on page 176.)

Introduction

For all the sea shanties about the "seven seas," a look at the globe reveals that there really is **only one ocean.** This vast ocean surrounds us all, no matter what continent we inhabit, and covers by far the largest proportion of surface area on Earth. It's thought to be the chemical crucible in which life itself began. The ocean—its nutrients and inhabitants, its awesome power, its tides and currents—has a tremendous impact on the planet and all life upon it. Awareness of the importance of the ocean, the dangers of pollution, and the increasing scarcity of its resources has grown markedly in recent years. The National Ocean Conference held in the summer of 1998 underlined these concerns.

Only One Ocean delves into major content areas in the earth sciences—especially oceanography, biology/life science, and environmental science. The guide consists of three main activities. Taken together, these activities:

- provide students with a grasp of the vast proportion of the Earth that is ocean, and an understanding that most of our valuable marine resources are found in relatively small areas of the ocean

- allow students to dissect and learn in detail about one fascinating open-ocean organism, the squid, in order to better understand animal characteristics and adaptations to a particular environment...as well as learn about the open-ocean environment itself

- introduce students to the significant and serious environmental issues presented by the increasing scarcity of food resources that humans derive from the ocean

- present the viewpoints of different interest groups seeking solutions to environmental problems

- encourage students to grapple with complex issues and decisions, and to share a sense of personal responsibility and stewardship for the well-being of our planet and the life it nurtures

This GEMS Teacher's Guide is adapted from the *MARE Teacher's Guide to the Open Ocean.* Drawn from the rich and varied collection of marine activities developed by the

MARE (Marine Activities, Resources & Education) program of the Lawrence Hall of Science, the activities in this guide bring together key science content with activity structures that promote language acquisition by all students. MARE (pronounced like the Latin word for ocean: mär´a) is an exciting whole-school, interdisciplinary, marine science program for elementary and middle schools (see the next several pages for more information). The MARE curriculum also emphasizes "Key Concepts" for each major activity—concepts not only communicated to the teacher, but directly conveyed to and displayed for the students. Two other MARE guides, *On Sandy Shores* (Grades 2–4) and *Ocean Currents* (Grades 5–8), have also been published by GEMS. (*Ocean Currents*, incidentally, makes an excellent companion guide to *Only One Ocean*.)

Key Concepts in *Only One Ocean* touch on the following:

- Most of our planet is covered by ocean, but **only a small fraction of the ocean supports large concentrations of life.**

- The investigation of the structure and function of an animal such as the squid can be used to study **adaptations** for an open-ocean ("pelagic") existence.

- Many people depend on squids for food or their livelihood. **More discussion among these people will help create solutions to the problem of diminishing squid populations.**

- Fish populations in the open ocean flourish where **the interaction of currents and sunlight provide a biologically productive environment.** Large commercial fisheries (industrial fishing operations) prosper in these locations.

- **Most of the ocean fisheries in the world are severely threatened** due to overfishing, bycatch (incidental capture of non-target species), or habitat loss.

- In order to **ensure that fisheries are sustainable,** public policy decisions and personal choices about what we buy and eat must be made **based on information about ocean systems.**

Activity-by-Activity Overview

In **Activity 1: Apples and Oceans,** students are introduced to the vastness of our planet's one, interconnected ocean; to the importance of the ocean to all life on Earth; and to the very limited resources we depend on from the land and sea. They begin by brainstorming what they know, value, and enjoy about the ocean. Pairs of students then use an apple and a circle graph (pie chart) to represent the planet in a very tangible way. They slice the apple and draw a corresponding chart in sections that illustrate various **critical resources** available from the land and ocean. Although a seemingly simple activity, the hands-on nature of apple-slicing and charting creates a striking visual of the state of our Earth's resources. This activity is a powerful model for changing student ideas about the productivity of the ocean and our role in protecting this valuable resource. Designing a mini-book, journal entry, or other creative writing piece later in the unit, students reflect on and demonstrate what they've learned.

In **Activity 2: Squids—Outside and Inside,** students palpably observe in depth (and inside!) one ecologically and economically important marine species—the squid. Squids are found throughout the planet's ocean, and are a vitally important food source for people around the world. Students begin by reading and discussing several statements about squids, and record their knowledge and any questions they have on individual charts. They share their prior knowledge with others—first with a partner, then in a foursome, then with the class.

Getting to the heart of this activity, students observe the squid's **adaptations** as they work in pairs to dissect a real squid. They investigate its structure and learn how all the parts function together to allow the squid to survive and thrive in its open-ocean environment. The dissection ends by honoring this marvelous organism in a squid feast, or Calamari Festival. The class next explores issues surrounding the threatened squid fishery by role-playing and discussing the problem from different points of view at a Squid Fishery Symposium. Students come up with their own solutions to an important global problem, setting the stage for the next activity.

In **Activity 3: What's the Catch?,** students call on what they've learned so far and acquire substantial new information on the ocean and its resources. They work cooperatively in small groups to learn about **fisheries,** the commercial fishing operations that are key players in global overexploitation of ocean resources. Students discover that most of the ocean fisheries in the world are severely threatened due to overfishing, bycatch (the incidental taking of species other than those targeted by the fishery), or habitat loss. In order to ensure that fisheries are sustainable, public policy decisions and personal choices about what we buy and eat must be based on information about ocean systems. The students present their findings to the rest of the class in a simulated World Fisheries Conference, in which they make recommendations to the nations of the world to help manage fisheries, and clarify their own personal decisions and choices.

Taken together, the three main activities in this unit convey important science content, allow students to investigate in detail the structures and adaptations of an intriguing open-ocean organism, and provide insight and information into major global environmental issues. The activities extend across the curriculum into language arts and social studies. They're as relevant as today's many news stories about the ocean and its resources, and they encourage students to think about their own personal opinions and responsibilities. The ocean, which spans the globe and makes up such a large proportion of our planet, provides an excellent way for students to really understand the global nature of many environmental issues. After experiencing this unit, your students will definitely have gained new insight into the environmental urgency of understanding that there is, indeed, "only one ocean."

The activities in *Only One Ocean* address the following key science standards and benchmarks outlined in the *National Science Education Standards* from the National Research Council, and the *Benchmarks for Science Literacy* from the American Association for the Advancement of Science (AAAS) Project 2061. The box that follows describes those standards and benchmarks and provides a teacher's overview of the major concepts and skills reinforced in the unit.

From the *National Science Education Standards:*

Science as Inquiry, Grades 5–8
Abilities necessary to do scientific inquiry
Understandings about scientific inquiry

- Students should develop the ability to listen to and respect explanations proposed by other students. They should remain open to and acknowledge different ideas, be able to accept the skepticism of others, and be able to recognize and analyze alternative explanations.

Life Science, Grades 5–8
Structure and function in living systems

- Living systems at all levels of organization—including cells, organs, tissues, organ systems, whole organisms, and ecosystems—demonstrate the complementary nature of structure and function.

- In multicellular organisms, specialized cells perform specialized functions, grouping and cooperating to form tissue, such as a muscle. Different tissues in turn group together to form larger functional units, called organs. Each type of cell, tissue, and organ has a distinct structure and set of functions that serve the organism as a whole.

Science in Personal and Social Perspectives, Grades 5–8
Populations, Resources, and Environments

- Causes of environmental degradation and resource depletion vary from region to region and from country to country.

From the *Benchmarks for Science Literacy:*

4B The Earth, Grades 6–8

- The Earth is mostly rock. Three-fourths of its surface is covered by a relatively thin layer of water (the oceans), some of it frozen. The entire planet is surrounded by a relatively thin blanket of air (the atmosphere).

- Earth's resources—fresh water, air, soil, trees, etc.—can be depleted by wasteful usage, or by deliberate or inadvertent destruction. The atmosphere and the oceans have a limited capacity to absorb wastes and recycle materials naturally. Cleaning up polluted air, water, or soil, or restoring depleted soil, forests, or fishing grounds, can be very difficult and costly.

5A Diversity of Life, Grades 6–8

- Animals and plants have a great variety of body plans and internal structures in order to make or find food, and to reproduce.

- All organisms, including humans, are part of—and depend on—interconnected global food webs. One of these food webs includes microscopic ocean plantlike organisms (phytoplankton), the animals that feed on them, and, finally, the animals that feed on those animals.

Into, Through, and Beyond

This GEMS Teacher's Guide features activities developed by the Marine Activities, Resources & Education (MARE) program at the Lawrence Hall of Science, and highlights several unique aspects of the MARE curriculum. Each activity in this guide, for example, is composed of three main pieces. These three sections are described in the "Overview" to each of the three main activities.

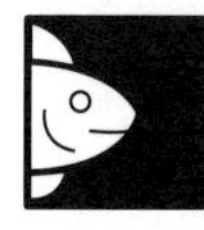

INTO THE ACTIVITY helps students recall how much they already know about the activity topic. Teachers may be pleasantly surprised by the wealth of knowledge students bring to the classroom, and may also become aware of significant misconceptions (or a lack of prior knowledge) that need to be addressed.

THROUGH THE ACTIVITY contains experiments, simulations, demonstrations, games, and facts to help students build on their prior knowledge and acquire, construct, and reflect on new concepts and information.

BEYOND THE ACTIVITY provides opportunities for students to explore the content further, applying what they have learned to new situations through self-designed projects, research, home activities, etc.

Activity Structures and Teaching Strategies

Several key "activity structures" developed by the MARE program are used in this guide. These are designed to help students talk, draw, and write about their related prior knowledge of a topic, or to distill and summarize what they've recently learned. These activity structures, with names like "Thought Swap" and "Think, Pair, Share," generally emphasize short, small-group discussions; cooperation; and social skills development. They create opportunities for students to use language in a nonthreatening and highly relevant setting in which the focus is on science content, not on the language itself. (In this regard these activity structures also support current research in second-language acquisition.) These activities are meant to be simple and accessible, and to help students simultaneously build their knowledge of science and their language skills. You may also find the activities help students develop their social skills as they begin working in cooperative groups.

In Culturally Diverse Classrooms

The activities in this guide were designed with both high-quality science content and language development in mind. The unit can be used with no modification in culturally diverse classrooms with large numbers of English-language learners. Students already fluent in English will, of course, enjoy the activities as well as the richness and depth of understanding that comes with integrating literature, discussion, art, video, journal writing, brainstorming, graphics, and cooperative projects with their study of science. The guide includes effective resources, assessments, and literature connections that provide teachers with tools to expand the unit.

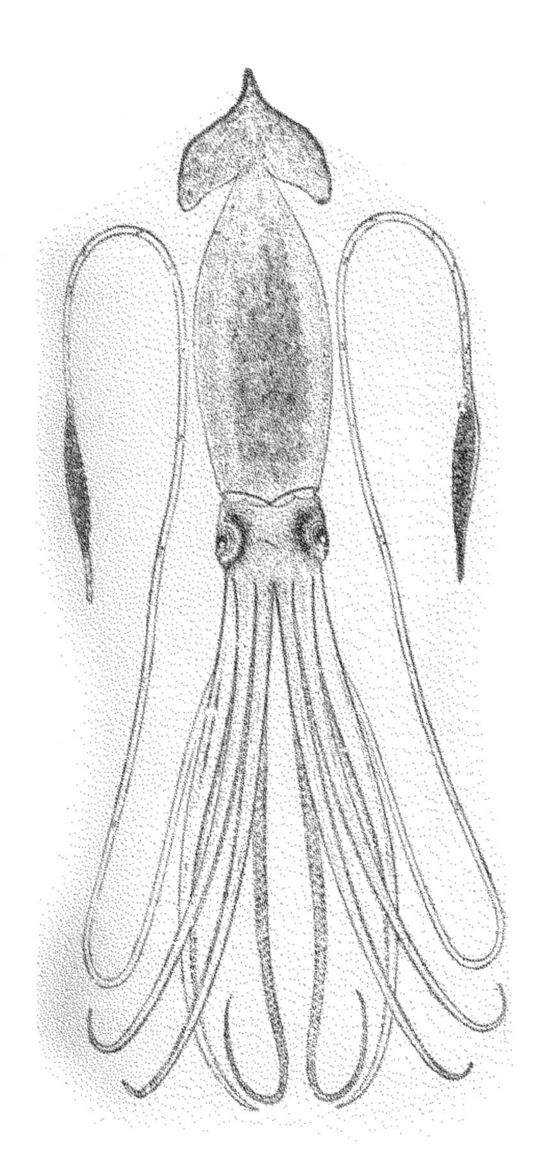

More About MARE

A program of the University of California at Berkeley's Lawrence Hall of Science, MARE is a yearlong, whole-school "ocean immersion" program. It engages parents, communities, and the entire staffs and student bodies of hundreds of schools in a comprehensive study of the ocean. In addition to providing curriculum, MARE offers teacher education in-services and summer institutes based on the most up-to-date scientific and educational research. The program focuses specifically on helping culturally, linguistically, and academically diverse schools to implement high-quality science education that is accessible to all students. Customized, whole-faculty in-services introduce teachers to new methods for developing their own integrated instructional plans based on the MARE curriculum, present marine science content, and help the entire school plan for its immersion activities. At MARE's two-week residential Summer Institute, teacher leaders sample hands-on activities, plan school-wide programs, learn from leading scientists and educators, and participate in exciting field experiences.

MARE transforms an entire school into a discovery laboratory for exploration of the ocean. The experience is often highlighted by a whole-school/whole-school-day "Ocean Month," an intensive educational event that creates an exciting atmosphere school-wide and serves as the centerpiece for yearlong ocean studies. Ocean Month builds a sense of inclusion throughout the school community and improves the general climate and educational culture of the school. Special-education, language-minority, and mainstream students work side by side across grade levels, peer-teaching and tackling special projects. Students have long, uninterrupted blocks of time to explore areas of interest in depth. Teachers receive on-site support from MARE staff, who coach, model-teach, coordinate, and dispense materials from MARE's extensive multimedia library. Parents are directly involved in the school's academic program.

The MARE curriculum focuses each grade on a different marine habitat and integrates language arts, language development, social studies, and fine art with science and mathematics. Many key science themes and concepts are explored. Integrating disciplines and linking subject areas, the curriculum helps students understand the overarching themes of science. Within each guide you will find in-depth teacher reference information, hands-on activities, teaching strategies, children's literature connections, planning materials for developing a comprehensive whole-school science program based on the study of the ocean, and suggestions for assembling student portfolios and conducting performance tasks to assess student achievement. Each activity in the MARE curriculum identifies and develops students' related prior knowledge through a rich variety of language experiences before introducing new topics. Each of the MARE guides can be used, at minimum, as a six- to eight-week science unit, or can be expanded and integrated into a comprehensive, yearlong science curriculum covering the disciplines of physical, biological, earth, and environmental sciences.

If you're interested in more information about MARE, please contact:

MARE
Lawrence Hall of Science
University of California
Berkeley, CA 94720-5200

phone: (510) 642-5008
fax: (510) 642-1055
e-mail: mare_lhs@uclink4.berkeley.edu

Web site: www.lhs.berkeley.edu:80/MARE/

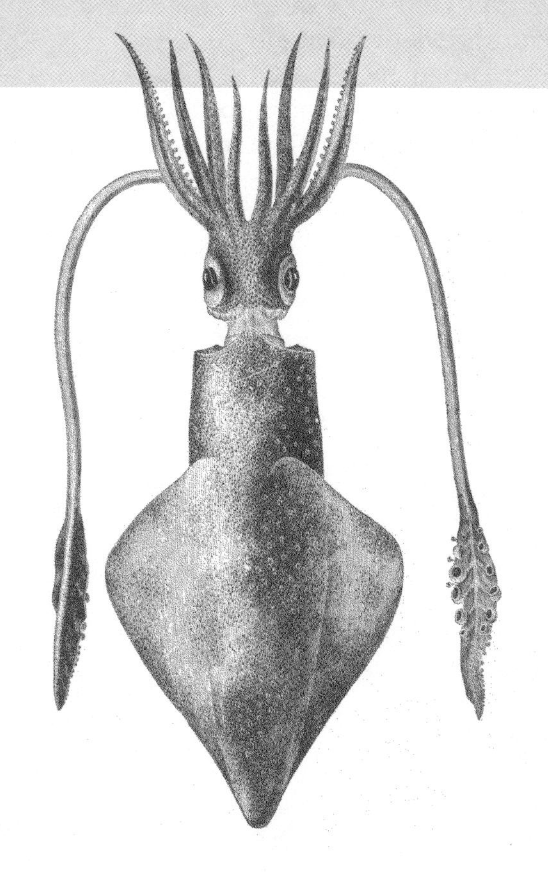

Time Frame

Depending on the age and experience of your students, the length of your class periods, and your teaching style, the time needed for this unit may vary. Try to build flexibility into your schedule so that you can extend the number of class sessions if necessary.

In Activity 1, Session 2, the option exists to spend an additional session or partial session presenting the subject of ***upwelling*** more thoroughly.

Activity 1: Apples and Oceans

Session 1: Brainstorm—Our "Planet Ocean" .... 45 minutes
Session 2: Apples and Oceans .... 45–60 minutes
Session 3: Creative Writing .... 45–120 minutes

Activity 2: Squids—Outside and Inside

Session 1: What about Squids? Anticipatory Charts ..... 45 minutes
Session 2: Hands-On Squid .... 45–120 minutes
Session 3: Debriefing Activities .... 45–60 minutes
Session 4: Squid Fishery Symposium .... 45–60 minutes

Activity 3: What's the Catch?

Session 1: Seafood Smorgasbord .... 45–60 minutes
Session 2: Fishery Experts .... 90–120 minutes (or 45–60 minutes plus homework)
Session 3: The World Fisheries Conference —Poster Presentations .... 45–120 minutes
Session 4: The World Fisheries Conference —Big Picture Recommendations .... 45–60 minutes

What You Need for the Whole Unit

The quantities below are based on a class size of 32 students. You may, of course, require different amounts for smaller or larger classes. This outline gives you a concise "shopping list" for the entire unit. Please refer to the "What You Need" and "Getting Ready" sections for each individual activity, which contain more specific information about the materials needed for the class and for each team of students.

A good deal of blank and lined paper is called for in this unit; we suggest you use the quantities listed below (under General Supplies) as a guideline, and that you have an extra supply on hand. (Much of the paperwork will be compiled as part of a class book, or to be used in student portfolios.)

Nonconsumables

- ❑ a videocassette recorder (VCR)
- ❑ a viewing monitor
- ❑ Monterey Bay Aquarium Video Collection: "Seasons of the Squid," or other video containing squid footage (see "Resources," page 163)
- ❑ 1 small jar with lid (like a baby-food jar) half-filled with water
- ❑ 1 "Fry Daddy" or other deep-fat fryer
- ❑ 1 medium bowl
- ❑ 1 spoon to mix the batter
- ❑ 1 knife to cut the squid
- ❑ 1 cutting board
- ❑ 17 hand lenses
- ❑ a large (36") inflatable globe (available in many travel stores and children's stores)
- ❑ 17 butter knives (metal preferred, or VERY sturdy plastic)
- ❑ 1 picture of each of the live animals from which the seafood for Activity 3 came
- ❑ pictures from magazines, old calendars, or postcards depicting the ocean or ocean life
- ❑ 1 Squid Interest-Group Profile student sheet (master on pages 85–86) cut into individual profiles
- ❑ 5 sets of 6 Fishery Information Cards (masters on pages 109–140)
- ❑ 5 sets of 6 Fishery Assignments (masters on pages 141–147)

Complete classroom kits for GEMS teacher's guides are available from Sargent-Welch. For further information call 1-800 727-4368 or visit www.sargentwelch.com

- ❑ *(optional)* 1 audiocassette player
- ❑ *(optional)* 1 audiocassette of natural ocean sounds
- ❑ *(optional)* laminated posters depicting physical or biological aspects of productive coastal zones (see "Resources")
- ❑ *(optional)* 1 microscope
- ❑ *(optional)* 1 small dish (such as a Petri dish) to use with the microscope
- ❑ *(optional)* 1 large world map
- ❑ *(optional)* 4–5 marine-life reference books (see "Resources")
- ❑ *(optional)* 1 world atlas
- ❑ *(optional)* book(s) of *haiku* poems

Consumables

- ❑ 17 apples (see "Getting Ready," page 17)
- ❑ 18 squids (see "Getting Ready," page 46)
- ❑ enough packages of tempura batter mix (the just-add-water kind) for the class (see page 71)
- ❑ 5 or more different seafood samples, enough for each student to have a small taste of each (see "Getting Ready," page 91)
- ❑ 50 full-sized, round, white, undivided, sturdy paper plates (Chinet® or similar)
- ❑ paper towels
- ❑ cooking oil for frying
- ❑ 1 box of toothpicks
- ❑ 1 "Land" poster or overhead transparency (example on page 17)
- ❑ 1 "Ocean" poster or overhead transparency (example on page 17)
- ❑ up to 45 sheets of chart paper (approximately 27" x 34") or blank overhead transparencies, to record brainstorming and display Key Concepts (chart paper preferred)
- ❑ 33 unlabeled External Squid Diagram student sheets (master on page 83) if you play Squid Jeopardy (see page 72)
- ❑ 33 unlabeled Internal Squid Diagram student sheets (master on page 84) if you play Squid Jeopardy
- ❑ 6 Squid Statements Anticipatory Chart student sheets (master on page 82)
- ❑ 1 Squid Statements Anticipatory Chart (example on page 82)
- ❑ 1 Squid Interest-Group Chart (example on page 48)
- ❑ 1 About Squids Chart (example on page 46)

- ❑ 3 squid anatomy charts: 1 "External," 1 "Internal," and 1 "What We Learned" (see Getting Ready, page 47)
- ❑ 5 copies of the World Map Handout (master on page 148)
- ❑ 5 copies of the Overview of the World's Ocean Fisheries handout (master on page 108), copied on colored paper
- ❑ 33 copies of the Dolphins and Your Tuna-Fish Sandwich handout (master on pages 105–107)
- ❑ *(optional)* 1 copy of the Squid Dissection Summary Outline for Teacher's Notes (master on pages 80–81)
- ❑ *(optional)* lemon for garnish

General Supplies

- ❑ a roll of masking tape
- ❑ 17 or more full sets of colored markers (wide-tipped, water-based)
- ❑ 5 or more full sets of fine-point markers
- ❑ notepaper for brainstorming
- ❑ 126 sheets of 8½" x 11" lined paper
- ❑ 33 sheets of 8½" x 11" blank paper
- ❑ 66 sheets of 10" x 14" or 11" x 17" blank paper
- ❑ lined paper or journals for creative writing
- ❑ 20 sheets of grid paper, for graphing assignments
- ❑ 33 pairs of scissors
- ❑ 17 rulers
- ❑ a 3-hole punch
- ❑ 3 brads for class book, *World Fisheries Conference Proceedings*
- ❑ at least 5 small paper clips
- ❑ pens or pencils for all students
- ❑ *(optional)* an overhead projector
- ❑ *(optional)* a good supply of blank overhead transparencies
- ❑ *(optional)* a laminator

Activity 1: Apples and Oceans

Overview

Our world is a water planet; nearly three-fourths of the Earth's surface is covered by ocean. Looking at a globe from the perspective of the vast Pacific Basin, it appears obvious that Planet Earth should more appropriately be named "Planet Ocean"!

When people say "the Earth," they often mean just the land portion of the planet. Of course Earth also means the entire planet—made up of land, ocean, and atmosphere. If we consider all of the planet, then much of it is in fact made up of "earth" (sand, rocky soil, mountains, the molten rock below the surface, mantle, core, etc.). But if we consider only the surface of the Earth, then the ocean represents by far the largest proportion.

The ocean allows life to exist. It makes our climate habitable, provides much of our oxygen and food, and transports—among much else—nutrients, people, and (sadly) even pollution around the globe. While the ocean may seem a limitless resource, only a small fraction is considered even moderately biologically productive. **Nearly all the marine resources we depend on come from a few small regions of the total ocean.**

In this activity, students are introduced to the vastness of our planet's one, interconnected ocean, and to the importance of the ocean to all life on Earth. They're also introduced to the concept of the very limited resources we depend on from the land (habitable area, farmable land, and fresh water) and from the ocean (upwelling areas that allow high biological productivity).

Benchmarks for Science Literacy *emphasizes the need for students to better understand and appreciate the global food web of which they are a part. The fragile resources of the ocean play a critical role in supporting life on our planet.*

In Session 1, students get "**Into** the Activity" by brainstorming what they know, value, and enjoy about the ocean. They go on to reflect on the benefits of the ocean to human life, and discover where most of life is found in the ocean. The class then goes "**Through** the Activity" over the next two sessions. In Session 2, students work in pairs, using an apple and a circle graph to represent the planet. To understand the intangible concept of limited land and water resources, students carefully section the apple and the graph into wedges representing various critical resources on the planet. These visuals give students an immediate sense of the small proportion of the Earth that provides resources from the land and the ocean.

To help students who may have trouble understanding fractional parts, a number line can be a helpful representation (see page 160).

In Session 3, students design a mini-book, journal entry, or other creative writing piece to demonstrate what they've learned. A number of "Going Further" activities are suggested for going "**Beyond** the Activity."

What You Need

For the class:

- ❑ a large (36") inflatable globe (available in many travel stores and children's stores)
- ❑ pictures from magazines, old calendars, or postcards depicting the ocean or ocean life
- ❑ up to 7 sheets of chart paper (approximately 27" x 34", to record brainstorming and display Key Concept
- ❑ 1 "Land" poster or overhead transparency (example on page 17)
- ❑ 1 "Ocean" poster or overhead transparency (example on page 17)
- ❑ a roll of masking tape
- ❑ *(optional)* an overhead projector (if you don't use chart paper for brainstorm)
- ❑ *(optional)* up to 7 blank overhead transparencies
- ❑ *(optional)* 1 audiocassette player
- ❑ *(optional)* 1 audiocassette of natural ocean sounds
- ❑ *(optional)* laminated posters depicting physical or biological aspects of productive coastal zones (see "Resources," page 163)
- ❑ *(optional)* 1 large world map

For each pair of students (plus one extra for the teacher):

- ❑ 1 apple
- ❑ 1 butter knife (metal preferred, or VERY sturdy plastic)
- ❑ 1 full-sized, round, white, undivided, sturdy paper plate (Chinet® or similar)
- ❑ 1 set of colored markers (wide-tipped, water-based)
- ❑ 1–2 paper towels

For each student:

- ❑ notepaper for brainstorming
- ❑ 1 sheet of 10" x 14" or 11" x 17" blank paper for mini-book
- ❑ lined paper or journals for creative writing
- ❑ a few fine-point markers
- ❑ 1 pair of scissors

Getting Ready

1. On chart paper, create two blank Brainstorm Charts (one for Session 1, one for Session 2). (You may wish to make a separate Question Chart as well, to record any questions that arise during the brainstorm.) Post the charts at the front of the room.

2. After reading the Apples and Oceans overview , read the Brief Introduction to Upwelling on page 32 and decide if you'll present this as a separate activity.

3. Have ready the large inflatable globe and the ocean posters (if you've decided to use them).

4. Make and hang the "Land" and "Ocean" posters at the front of the room where you can easily write on them, or create and set up the transparencies on the overhead projector. (It's a nice touch to add small drawings, as shown in the illustration below.)

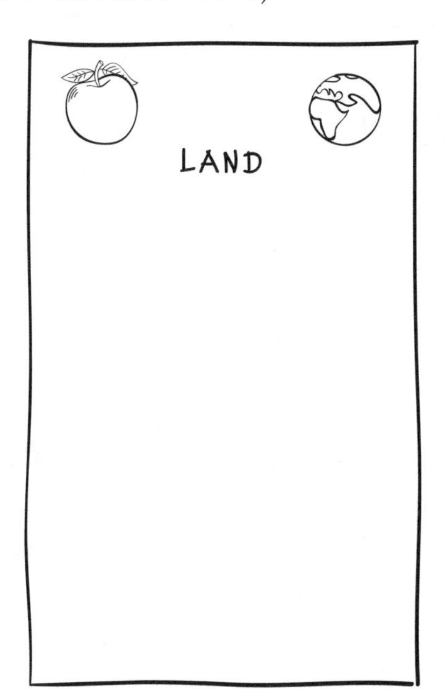

5. Write out the Key Concept for this activity in large, bold letters on a half-sheet of chart paper, and set aside:

- **Most of our planet is covered by ocean, but only a small fraction of the ocean supports large concentrations of life.**

6. Buy and wash the apples (students may want to eat them at the end). We recommend soft apples like red delicious or golden delicious (not pippins) that will cut easily.

Using apples to represent our planet is an effective and hands-on way for students to get a sense of the limited resources our world supports. At the end of the activity, if students don't wish to eat the apples, have them brainstorm ways in which they can be recycled or reused.

To better understand "where you're going" with this activity, see the completed circle graph on page 30.

ABOUT THE OCEAN

a. What do you know and enjoy about the ocean?
b. How do people depend on the ocean?
c. What are the things people do that affect the ocean and the life that lives there?
d. What areas of the ocean do you think contain the most life?
e. Where do you think the best spots are in the ocean to catch fish?

7. Read through all the sessions, especially Session 2: Apples and Oceans. In this session, you'll guide your students through a series of apple slicing and charting activities. The steps may seem a bit stiff at first, but they outline the "flow" of slicing and recording. The process will move quickly as you become accustomed to the back-and-forth nature of the activity. As the slicing progresses, your students will be amazed at the tiny pieces that represent fresh water and habitable land.

8. If you've decided to use them, set up the ocean-sounds audiocassette and cassette player. Gather the ocean pictures and photos for the brainstorm. Write the following questions on chart paper, the board, or an overhead transparency:

a. What do you know and enjoy about the ocean?

b. How do people depend on the ocean?

c. What are the things people do that affect the ocean and the life that lives there?

d. What areas of the ocean do you think contain the most life?

e. Where do you think the best spots are in the ocean to catch fish?

9. Decide how you'll divide students into the pairs and small groups needed for this activity, and have all materials at the ready.

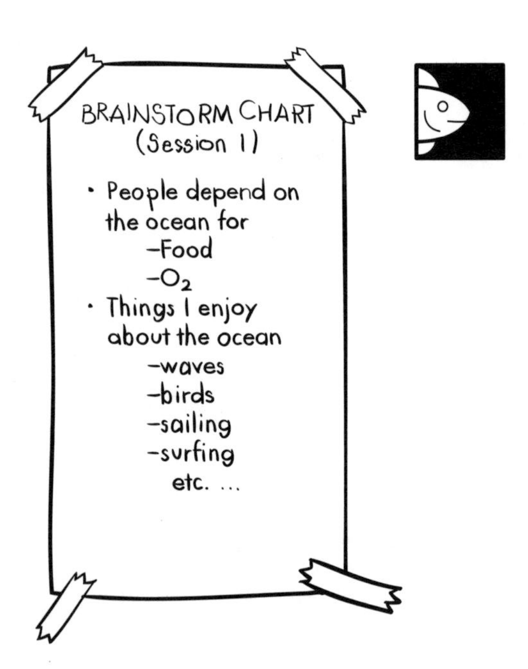

Session 1: Brainstorm—Our "Planet Ocean"

1. Tell the class they're about to study the ocean. Explain that when you say "the ocean," you mean the planet's one, vast ocean—not just the Atlantic, Pacific, Indian, etc. Arrange students in groups of four or six. Pass out the pictures and photos and play the ocean-sounds cassette. Ask students to discuss the questions you've posted, encouraging them to use the pictures to help them brainstorm.

2. Tell them that as they talk, they might come up with questions about the ocean. Have them jot these down on paper for later reference. Circulate among the groups to encourage discussion and to listen to their ideas.

3. After a few minutes, invite the groups to share their brainstorm ideas and questions. Refer back to the original

posted questions to get the sharing going. Make sure each group gets a chance to share a few thoughts or questions, and record these on the Brainstorm Chart for Session 1.

4. Pull out the globe and show the class the "traditional map view" of the world—that is, with the continents in full view with the Americas in the center. Ask the students what they can tell about the world from this perspective. [Big continents of land surrounded by water.] Now, turn the globe to show the "Pacific Ocean view." What does *this* view tell about the world? [Most of the Earth is covered by ocean!]

The ocean was the blue part of the globe with lines and letters on it.

—from "Earliest Memories of the Ocean," MARE Summer Institute, 1994

5. Lead a brief discussion, based on this new perspective and the ideas brought up in the brainstorm, to introduce these important concepts:

- Most of our planet is covered by ocean.
- About 95 percent of all *living space* on the planet (for all organisms) is located in the ocean.
- About 85 percent of everything that lives on Earth lives in the ocean.
- About 70 percent of the oxygen we breathe comes from the process of photosynthesis by microscopic plantlike organisms, or ***phytoplankton.***
- People get food and water from the ocean.
- The ocean plays a major role in moderating our climate. Without an ocean, the surface of our planet would freeze at night and be too hot for most life to endure during the day.

The vast expanse of ocean water is less changeable than land; it acts as a heat "sink" at night, keeping the planet from freezing. During the day, in certain areas, winds sweeping over the ocean pick up air evaporated from cold ocean currents and carry it over the land, keeping the planet cool.

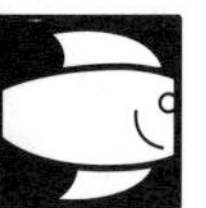

Session 2: Apples and Oceans

1. Tell students they're going to discover even more about the critical resources we depend on from "Planet Ocean." Hold up an apple and explain how they'll use this model to better understand the limited resources available on our planet. To help students see the activity as more than just an apple-cutting experience, tell them scientists often use a smaller model to help explain ideas about something too large to work with—in this case, our planet!

2. Have students clear all materials off their tables and sit side by side with a partner. Pass out to each pair an apple, a knife, a paper plate, a paper towel, and a set of colored markers. Keep a complete set for yourself to demonstrate with. Ask students to be careful with the knives, and **not to make any cut until *after* you've demonstrated it.**

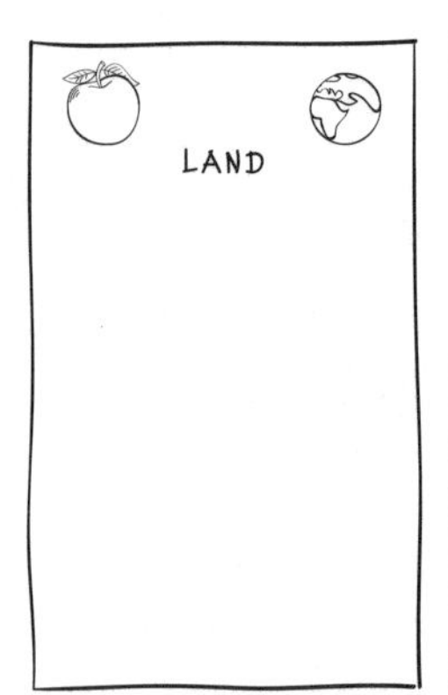

3. Hold up an apple and the globe, point to the "Land" and "Ocean" posters or transparencies, and again tell the students they'll be using the apple as a model to represent the planet.

4. Have them quickly decide (or you can assign) which partner will be the **"land"** person and which will be the **"ocean"** person. (Let them know they'll switch jobs half-way through the activity.) Have the land people raise their hands, and tell them that they should start with the apple, the knife, and a paper towel to cut on.

5. Have the ocean people raise their hands, and tell them that they should start with the plate and the markers. Explain that the land people are going to make some cuts on the apple, and the ocean people will make a circle graph to represent each of those cuts.

The content you'll present with the cutting of the apple is fairly complex and moves fairly quickly, so it's especially important for English-language learners that you establish a consistent "rhythm" with the sequence. Students will soon become aware that there are several opportunities to understand each "fact" (see you cut it, cut it themselves, hear you say it, see you and their partner graph it, see you write it on the poster). This should help lower their anxiety.

6. Explain that you'll demonstrate each step before students do it themselves. Students will hold up their apple slices and graphs after each step, so that you and their partners can check them for understanding.

7. Again, show the "Land" and "Ocean" posters at the front of the room or on the overhead projector. Tell the students you'll also record information here for everyone to see.

7th- and 8th-grade teachers have suggested eliminating the teacher demonstration of the circle graph throughout the lesson, to allow students to do it on their own without the prompts. (Read ahead to eliminate your involvement in this step if that's what you decide.) Alternatively, teachers have used the circle graph activity as a separate assessment activity following this lesson.

For each slice of the apple, you'll go through the following sequence:

1. Teacher gives a direction about how to cut the apple.

2. Teacher cuts the apple to demonstrate.

3. "Land people" students cut the apple and hold up the piece. Teacher asks students what part of the planet this represents.

4. Teacher explains the significance of the fractional part of the planet this slice represents and, when applicable, shows its corresponding part on the globe.

5. Teacher draws a corresponding section on the circle graph to demonstrate. (This step is optional—we suggest you read both sidebars before deciding.)

6. "Ocean people" students draw a corresponding section on their circle graphs and hold them up.

7. Teacher writes and labels the fraction on the poster or transparency.

Slice One: $\frac{1}{4}$ of the Planet is LAND

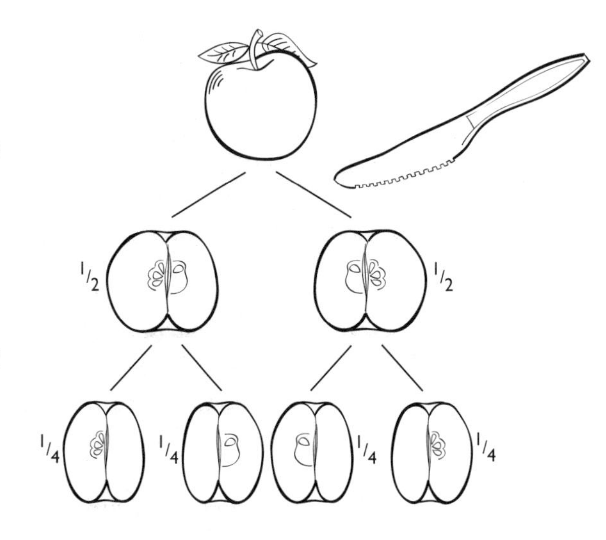

1. **Teacher gives direction and cuts the apple.**
Tell the land people that they're going to cut the apple into four equal pieces by cutting it in half through the core and cutting each of those halves in half again. Cut your apple to demonstrate.

2. **Students cut the apple.**
Have students cut their apples and hold up one of the four pieces when they're finished. Ask them, "What fractional portion of the planet does this represent?" [$\frac{1}{4}$.]

3. **Teacher explains.**
Explain that this section of apple represents $\frac{1}{4}$ of the planet, the portion that is covered by *land*. Show this on the globe. The other three sections represent the $\frac{3}{4}$ of the world covered by *ocean*; have students set those three pieces aside for the ocean part of the activity.

You may wish to have students place their apple slices on a piece of paper, after they've cut them, and label each section (fractional part) to refer to later.

4. **Teacher draws on the circle graph.**

   a. Tell the ocean people they're going use their plates to make a circle graph representing the ocean and land portions of the planet. Ask students how to divide the plate into these fractional parts.

   b. Demonstrate as necessary, holding up your plate and drawing with a colored marker. First draw a line (diameter) from one point on the circumference of the plate to a point on the opposite side (going through the center point), to divide the plate in half. Draw a second diameter, perpendicular to the first, to divide the plate into fourths.

   c. Model one way to label the circle graph: On the rim of the plate, label each of three fourths "ocean" and the remaining fourth "land."

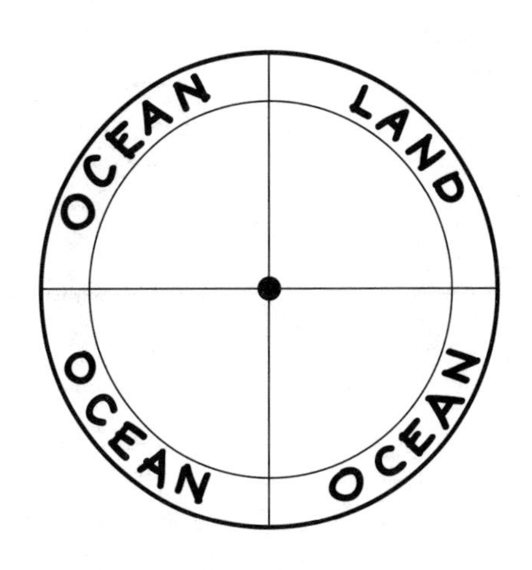

5. **Students draw on the circle graph.**
Have students create their graphs. Circulate among them and assist any students who are having difficulty.

6. **Teacher writes on the poster or transparency.**
At the top of the "Land" poster or transparency, under LAND, write "$\frac{1}{4}$ Land."

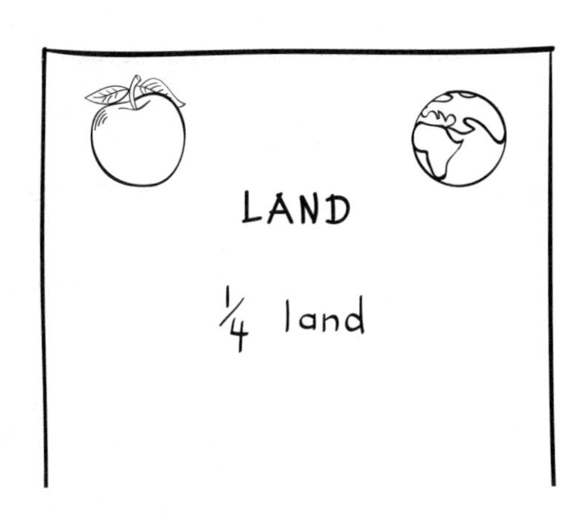

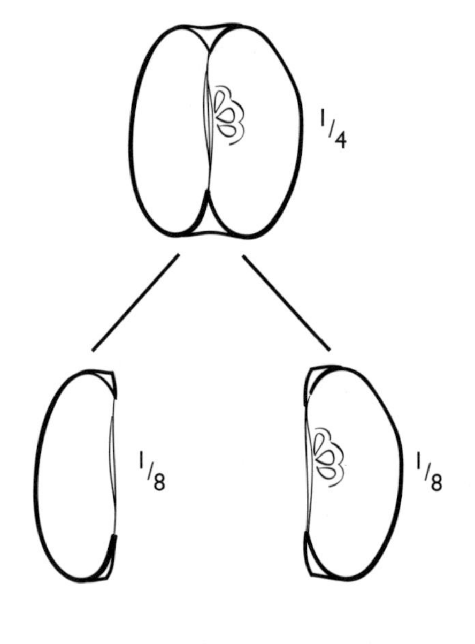

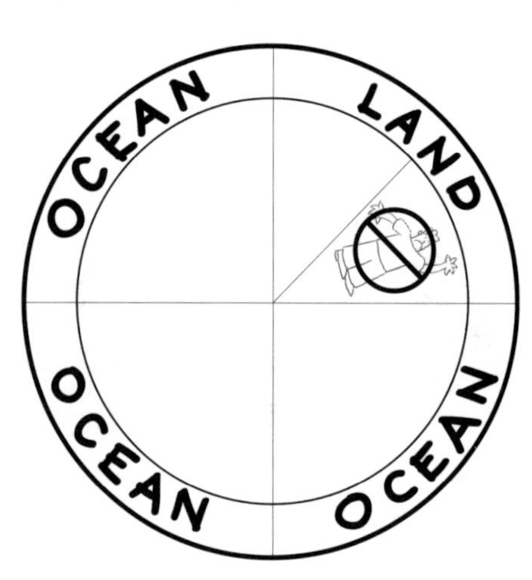

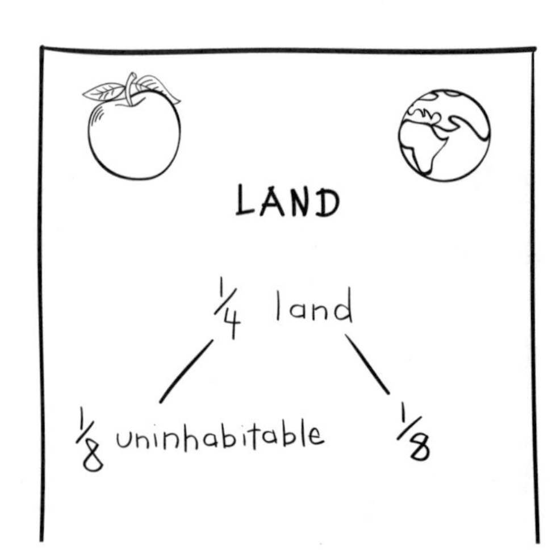

Slice Two: $\frac{1}{8}$ of the Planet is Uninhabitable Land

1. **Teacher gives direction and cuts the apple.**
Tell the land people they're going to cut the piece of the apple representing the land into two equal pieces. Divide your $\frac{1}{4}$ piece, either lengthwise or across the widest section. Hold up the two slices and demonstrate that they're equal.

2. **Students cut the apple.**
Have students cut their apple fourths into two equal pieces and hold up one of the resulting slices. Ask them, "What fractional portion of the planet does this represent?" [$\frac{1}{8}$.]

3. **Teacher explains.**
Explain that this piece represents all the land on Earth that's too dry, too wet, too cold, or too hot for people to live on. This is *uninhabitable* land, such as many mountain tops, river basins, deserts, etc. Show a few examples on the globe.

4. **Teacher draws on the circle graph.**

a. Tell the ocean people they're going to draw the uninhabitable portion of the land on the circle graph. Have them locate the land on their plates. How might they divide it?

b. Demonstrate how to draw a line (radius) from the center point of the plate to the rim, dividing the $\frac{1}{4}$ land into two equal pieces.

c. Model how to label one of the $\frac{1}{8}$ portions of the circle graph with a picture of a frowning person inside a red circle with a line through it. This represents the "uninhabitable" portion of the land.

5. **Students draw on the circle graph.**
Have students record on their graphs. Walk around holding yours up and assist students as necessary.

6. **Teacher writes on the poster or transparency.**
On the poster or transparency, under $\frac{1}{4}$ Land add a branch labeled "$\frac{1}{8}$ Uninhabitable."

Slice Three: $\frac{1}{8}$ of the Planet is Habitable Land

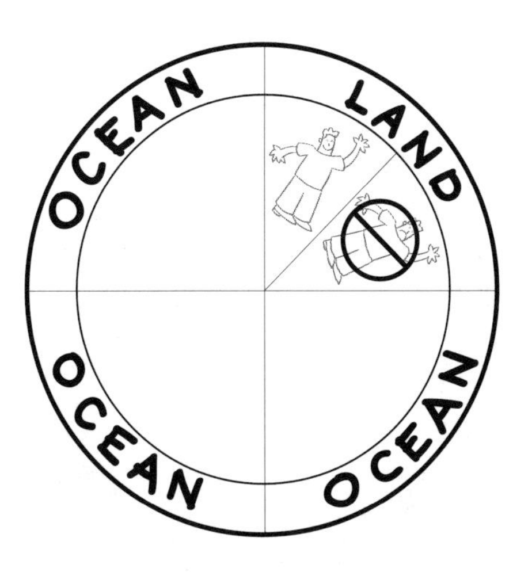

1. **Teacher explains and draws on the circle graph.**
Tell land students to follow along as you set aside the $\frac{1}{8}$ of your apple representing the uninhabitable land. Hold up the remaining $\frac{1}{8}$ piece of apple and have the students do the same. Ask them, "What fractional portion of the planet does this represent?" Explain that it represents the *habitable* land (where people *can* live). Show a few examples on the globe. Now hold up your plate and label this $\frac{1}{8}$ of the graph "habitable" by drawing a picture of a person smiling.

2. **Students draw on the circle graph.**
Have ocean people do the same on their graphs.

3. **Teacher writes on the poster or transparency.**
On the poster or transparency, next to $\frac{1}{8}$ Uninhabitable add a branch labeled "$\frac{1}{8}$ Habitable."

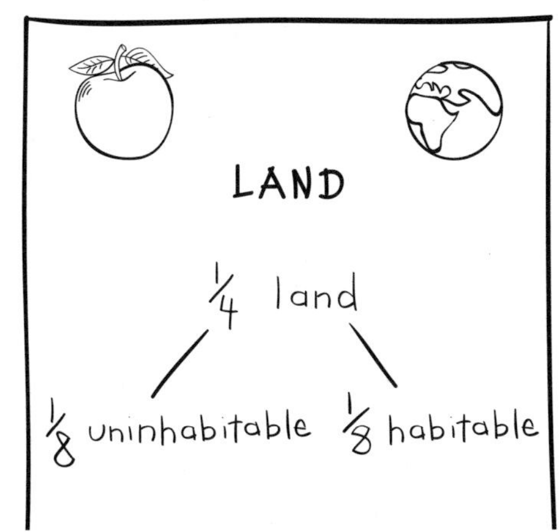

Slice Four: $\frac{1}{32}$ of the Planet is Farmable Land

1. **Teacher gives direction and cuts the apple.**
Demonstrate how to divide the $\frac{1}{8}$ piece representing the habitable land into four equal pieces, as follows:

    a. Cut the $\frac{1}{8}$ section into two equal pieces. Hold up one of these pieces and ask students, "What fractional portion of the planet does this represent? [$\frac{1}{16}$.]

    b. Still holding up the $\frac{1}{16}$ piece, ask, "If we cut *this* piece into two equal pieces, what fractional part will we have? [$\frac{1}{32}$.] Cut the $\frac{1}{16}$ section into two equal pieces. Repeat with the other $\frac{1}{16}$ section.

2. **Students cut the apple.**
Have students divide their $\frac{1}{8}$ piece into four equal pieces, as you did. As they hold up one of these pieces after cutting the apple, ask them, "What fractional portion of the planet does this represent?" [$\frac{1}{32}$.]

3. **Teacher explains.**
This small fractional piece represents all the farmable land that grows food for people, as well as all the pasture and graze land that grows food for animals that then become food for people. Alarmingly, much of our farmable land in the U.S. is paved over each year to build towns and highways. On the globe, show a few examples of existing farmable land, and of land that was farmable in the past but has since been developed.

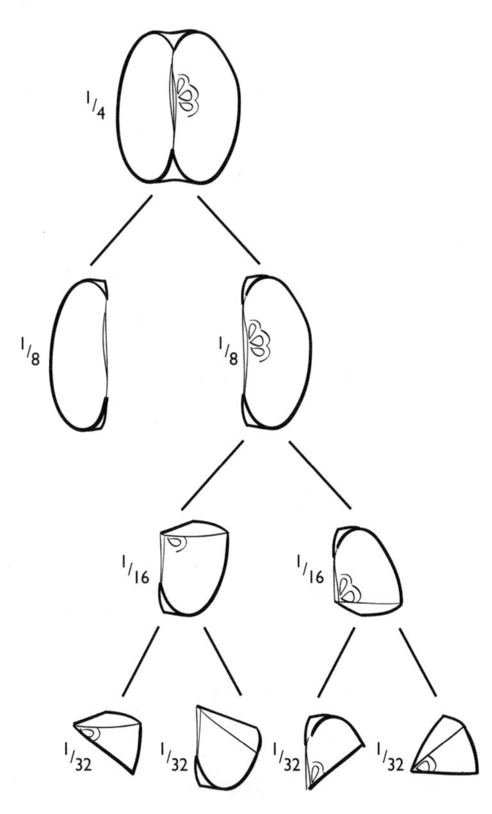

Most atlases can point you to specific regions where human development has altered the look and usability of the land over time.

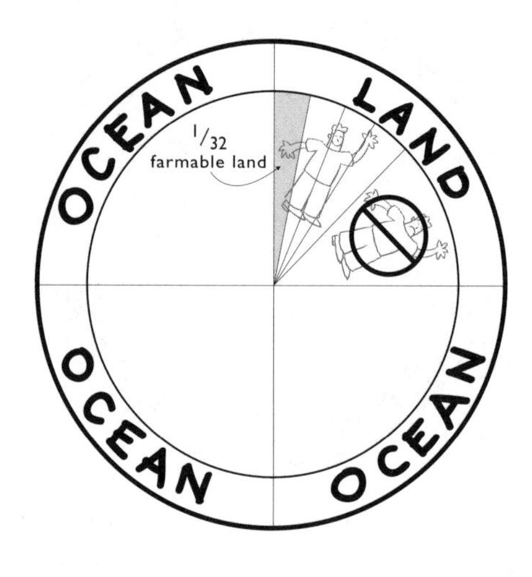

4. **Teacher draws on the circle graph.**
Divide the $\frac{1}{8}$ wedge representing the habitable land on the paper plate in half, then divide each of the resulting $\frac{1}{16}$ pieces in half again. Color one of the resulting $\frac{1}{32}$ pieces (ideally, green) to represent the habitable land on which we can grow food. With an arrow, label this colored slice "$\frac{1}{32}$ Farmable Land."

5. **Students draw on the circle graph.**

6. **Teacher writes on the poster or transparency.**
On the poster or transparency, under $\frac{1}{8}$ Habitable, add two branches labeled $\frac{1}{16}$. Under one of the $\frac{1}{16}$ branches, add two $\frac{1}{32}$ branches, and label one of them "Farmable Land."

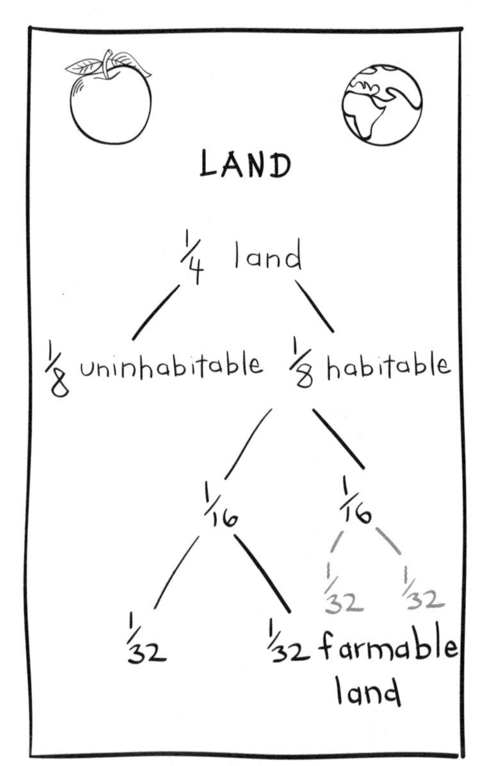

Slice Five: $\frac{3}{10,000}$ of the Planet is Land with Drinkable Water

1. **Teacher gives direction and cuts the apple.**
Take the $\frac{1}{32}$ piece of the apple and (carefully!) cut off the thinnest, tiniest sliver possible and hold it up.

2. **Students cut the apple.**
Have students hold up their tiny slivers after cutting and ask them, "What fractional portion of the planet does this represent?" [It's so tiny that it's impossible to tell!]

3. **Teacher explains.**
This tiny sliver (still too big to reflect the actual proportions) is just a model that represents $\frac{3}{10,000}$ of the Earth's surface. This area supplies all the drinkable water on the planet (including both habitable and uninhabitable parts)! It represents all of our lakes, ponds, rivers, streams, reservoirs, and underground aquifers—all of the fresh water that is accessible to us. All life on land—every plant and animal, including every human—depends on fresh water for survival…and look how little of it there is! Show a few examples on the globe. Briefly discuss such water conservation issues as drought, pollution, water diversions, water use and waste, etc. Have students be sure not to lose this very important slice—you'll want to look at it again later.

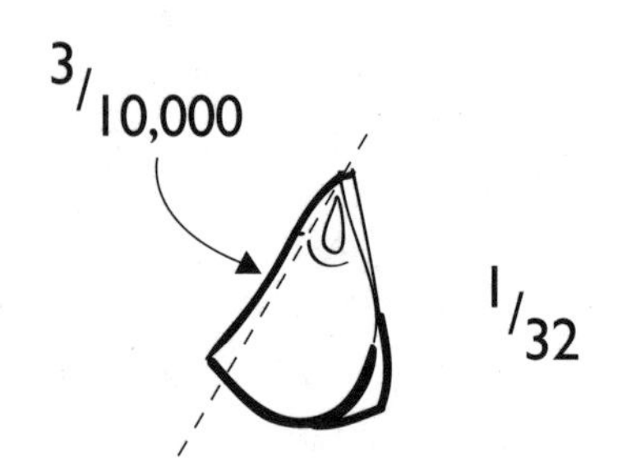

4. **Teacher draws on the circle graph.**
Demonstrate how to represent the drinkable water on the plate using a few small dots (of a single color) in each of the land sections, both habitable and uninhabitable. (The number of dots doesn't matter.) Label these dots "$^{3}/_{10,000}$ Drinkable Water."

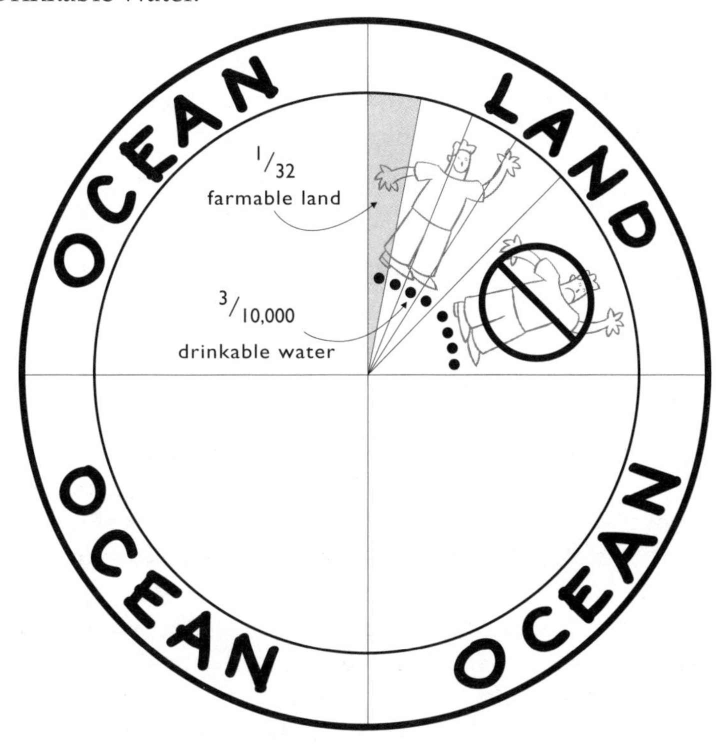

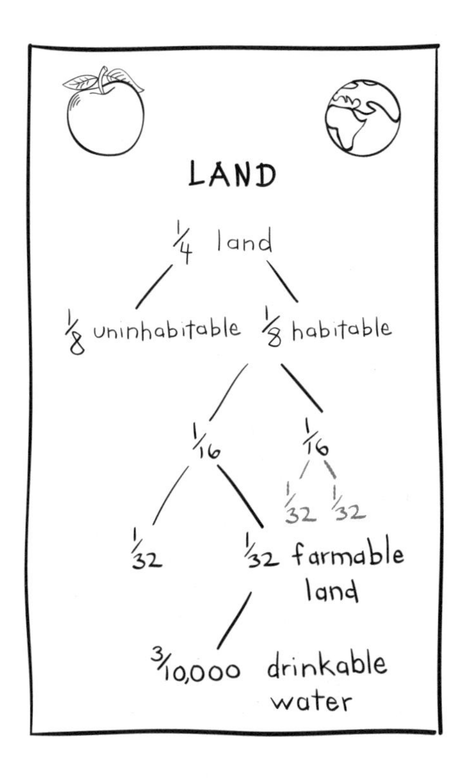

5. **Students draw on the circle graph.**
Have students record these dots on their graphs.

6. **Teacher writes on the poster or transparency.**
On the poster or transparency, under $^{1}/_{32}$ Farmable Land, add a branch labeled "$^{3}/_{10,000}$ Drinkable Water."

Slices Six and Seven:
$^{3}/_{4}$ of the Planet is OCEAN

Before continuing: Tell the class that now that they've seen how little fresh water there is on our planet, they're going to focus again on the ocean. Set aside the "land pieces" of the apple **(save the "drinkable water"!),** and return to the $^{3}/_{4}$ of the apple representing the ocean. **Have the partners switch jobs** so that the ocean partner is now cutting the apple and the land partner is now drawing on the paper plate. Call everyone's attention to the "Ocean" poster.

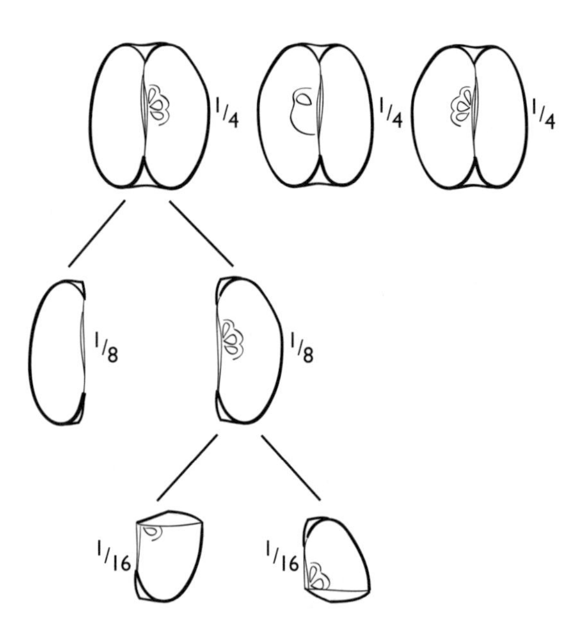

Although we think of the ocean as a vast, infinite resource, most regions of the world's ocean are not very productive; there's little life in these regions, and we consider them biological deserts. This little slice of apple represents the small region of the ocean considered even moderately productive. This area of concentrated ocean productivity is found over the shallow parts of the continental shelf. The shelf usually ends at 200 meters (660 feet) in depth and, on average, about 80 kilometers (50 miles) from shore (closer on the west coast of North America, farther on the east). The area of the planet's coastal zone is about four times the area of the lower 48 United States.

Slice Six: 1/16 of the Planet is Productive Coastal Zone

1. **Teacher gives direction and cuts the apple.**
Tell the ocean people that they'll cut one of the three fourths that represents a portion of the ocean in half. Ask, "What fractional parts will those two pieces be?" [1/8.] Demonstrate the cut and hold up one of the 1/8 pieces. Take that piece and cut it into two equal pieces. Hold up one of these small pieces and ask, "What fractional part of the planet does this represent?" [1/16.]

2. **Students cut the apple.**
Have students cut their apples and hold up one of the small pieces when they're finished. Review: "What fractional portion of the planet does this represent?" [1/16.]

3. **Teacher explains.**
Ask, "What kinds of things do we eat from the ocean?" [Fish, shrimp, clams, seaweed, etc.] "Where in the ocean do you think we harvest (catch) the most fish, shrimp, or clams?" Explain that the ***fisheries*** (commercial fishing operations) are located in these areas. This piece, 1/16 of the planet's surface, approximately represents the biologically productive ***coastal zones*** of the ocean, where almost all (90%) of the world's fisheries occur. Looking at this fractional part, you can see that only a tiny portion of the ocean produces most of the seafood we eat! Show a couple of examples on the globe—perhaps off the west coasts of North America and Africa.

4. **Teacher draws on the circle graph.**
Demonstrate for the land people how to represent the productive coastal zone on the circle graph, before they follow along on theirs. Divide one of the fourths marked "ocean" in half. Then draw another line dividing one of these eighths in half to create sixteenths. Label one of the 1/16 sections the "Productive Coastal Zone" and draw a fish or squid in it.

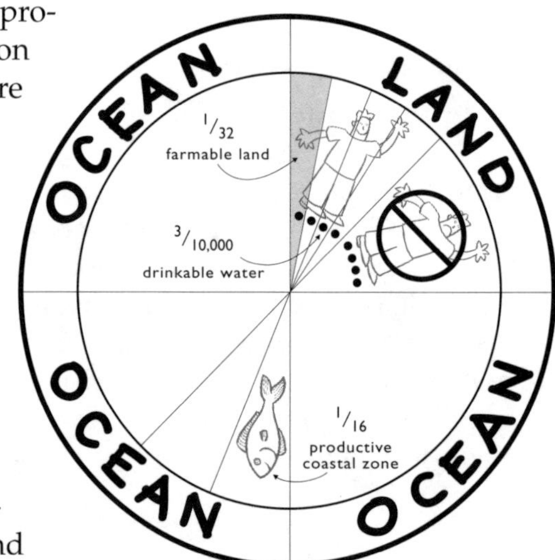

5. **Students draw on the circle graph.**
Have land people record the productive coastal zone on their graphs. Circulate among the students, holding up your graph and assisting as needed.

6. **Teacher writes on the poster or transparency.**
At the top of the "Ocean" poster or transparency, under OCEAN, write "$\frac{3}{4}$ Ocean." Under that, add branches for $\frac{1}{4}$, $\frac{1}{8}$, and $\frac{1}{16}$ as shown in the illustration. Label one of the $\frac{1}{16}$ branches "Productive Coastal Zone."

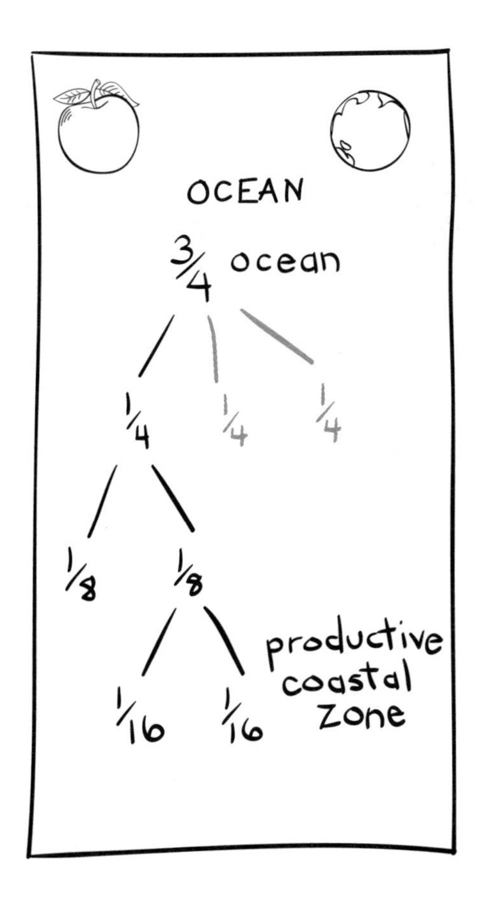

Slice Seven: $\frac{3}{4,000}$ of the Planet Is Upwelling Zone

1. **Teacher gives direction and cuts the apple.**
Take the $\frac{1}{16}$ piece of the apple. Cut off a very thin sliver and hold it up.

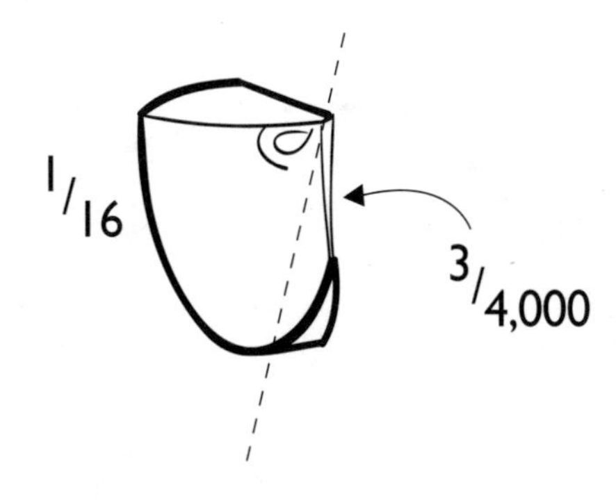

2. **Students cut the apple.**
Have students cut a sliver off one of their $\frac{1}{16}$ pieces. As they hold them up, ask, "What fractional portion of the planet does this represent?" [Again, it's so tiny that it's impossible to tell!] Ask them not to lose this sliver.

3. **Teacher explains.**
This tiny slice (which, like the drinking-water slice, is still too big to be anything but representative) is a model that represents $\frac{3}{4,000}$ of the world's surface. It represents the six tremendously productive ***upwelling*** areas found within the coastal zone. Explain that upwelling is a process that brings very cold, nutrient-rich water from deep down in the ocean up to the surface during some seasons on the west coasts of six continents. Show a few examples on the globe (see number 4, below). The highest concentrations of productivity are found in upwelling areas—these, as well as the polar areas in summer, are by far the most productive areas of the world's ocean. They're the prime destinations for migrating birds and marine mammals such as seals and whales.

For more information on upwelling, or to add a special "Going Further" session on this topic, see page 32 following this session.

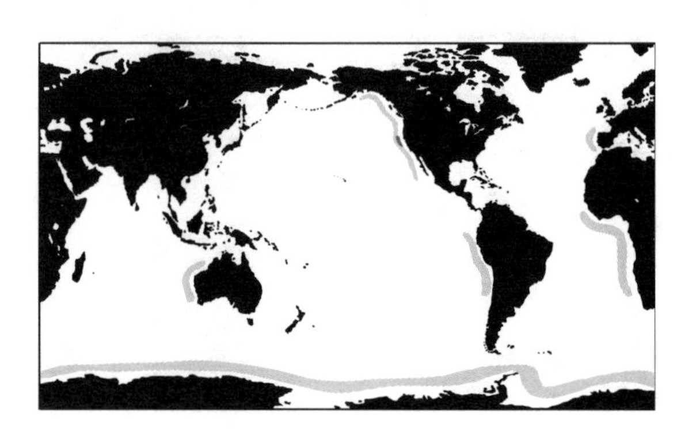

4. **Teacher draws on the globe.**
On the globe (using a water-based marker) draw a narrow band along the Pacific coast of North America to illustrate how small this upwelling zone is. Highlight the Central California coast as an even smaller subset. Mark the other five major upwelling areas on the planet: along the central and southwest coast of Africa; the central west coast of South America; the west coast of Australia; the west coast of Europe (Spain and Portugal); and around Antarctica. We're responsible for caring for and sharing this huge and vital world resource.

One of these upwelling areas is found along the Pacific coast of North America, making that stretch one of the biologically richest regions of the ocean. Roughly one-fourth of the entire world's upwelling zones lie off the west coast of North America between Canada and Baja California, Mexico. Along that coast, the Central California portion is by far the most biologically productive.

5. **Teacher draws on the circle graph.**
Demonstrate how to represent the rich upwelling zones on the circle graph, using six small dots (of another single color) within the productive coastal zone. With an arrow, label them "$^3/_{4,000}$ Six Upwelling Zones." Your circle graph is now complete.

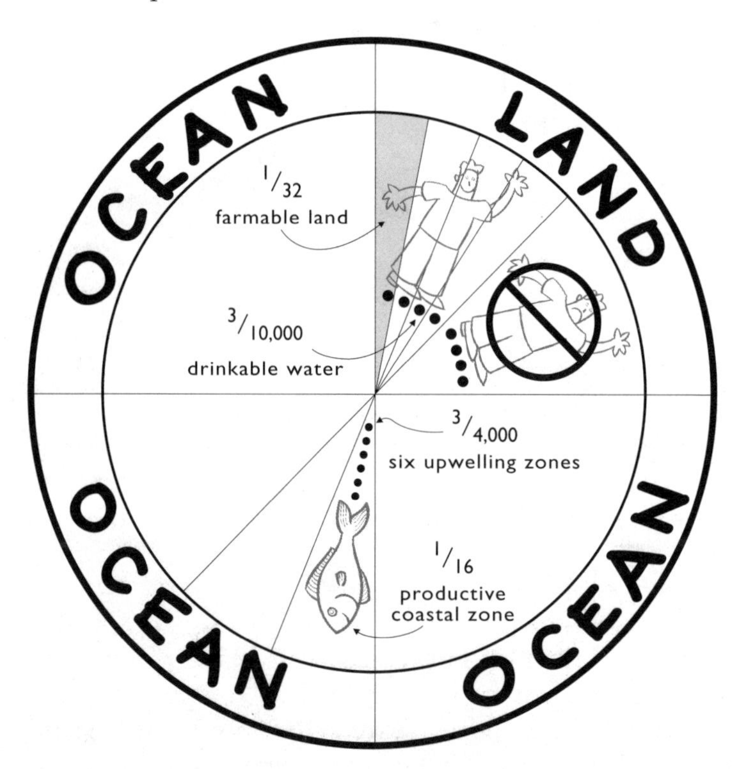

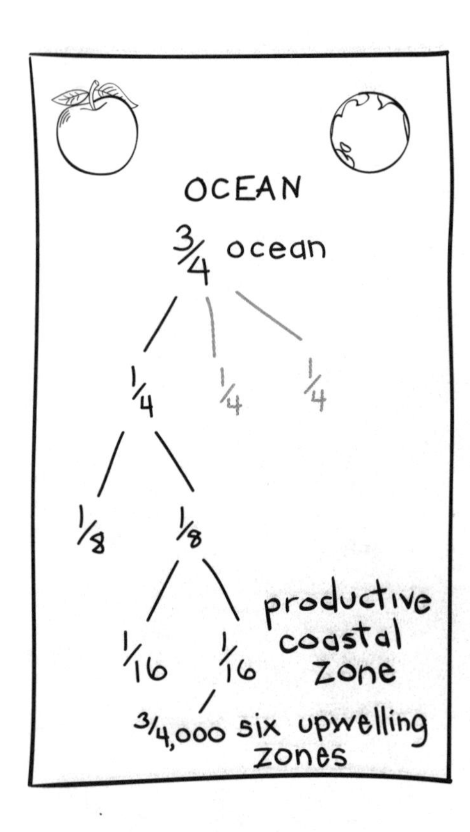

6. **Students draw on the circle graph.**

7. **Teacher writes on the poster or transparency.**
On the poster or transparency, under $^1/_{16}$ Productive Coastal Zone, add a branch labeled "$^3/_{4,000}$—Six Upwelling Zones."

Note: At this point, some teachers like to introduce the concept of the **photic zone,** as follows: Peel off a piece of the apple skin to represent part of the ocean, and hold it up. This piece of skin, though it's too thick to be truly accurate, represents the **photic zone**—the top 100 meters (330 feet) through which light can penetrate and support photosynthesis for microscopic plantlike organisms (phytoplankton). Since these organisms form the base of the ocean food pyramid, **all** life in the ocean depends on the existence of a productive photic zone. Show the photic zone of the ocean on the graph as a (single-colored) dot or two in each of the three sections of the ocean (not shown in circle graph illustration). Label these dots: "Photic Zone." (See sidebar.)

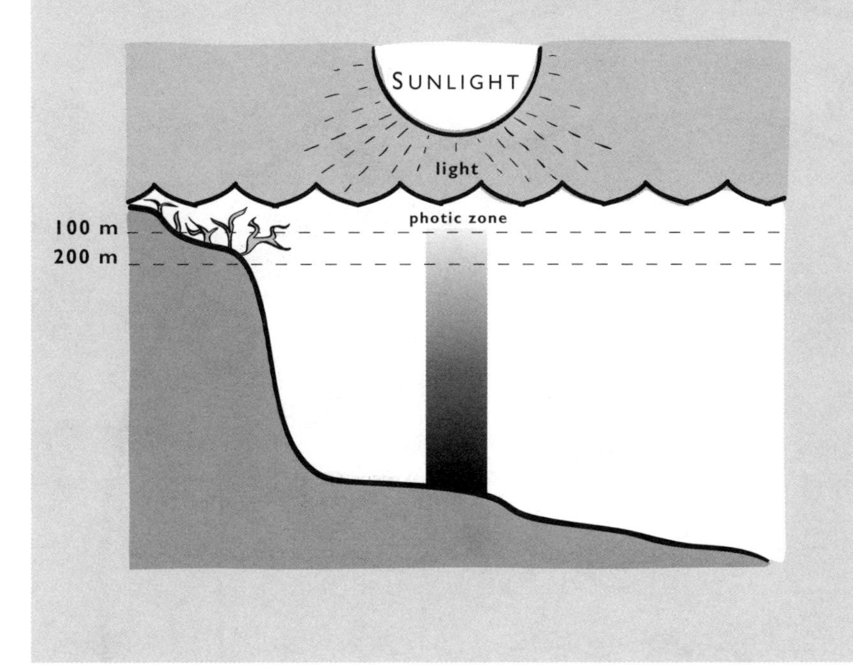

Almost all life in the ocean is concentrated in the shallow region just below the surface in a narrow band along the coasts. *Photosynthesis occurs across the entire ocean—but near the coasts, where the water is relatively shallow, the photic zone can extend all the way to the ocean floor. This provides enough sunlight for large seaweed—which anchors on the sea floor—to photosynthesize and grow. It's also along the coastlines that sunlight, carbon dioxide and oxygen, and sufficient nutrients (both from land runoff and from upwelling) combine to provide all the ingredients for phytoplankton to thrive. The rate of photosynthesis—and therefore productivity—is greatest here.*

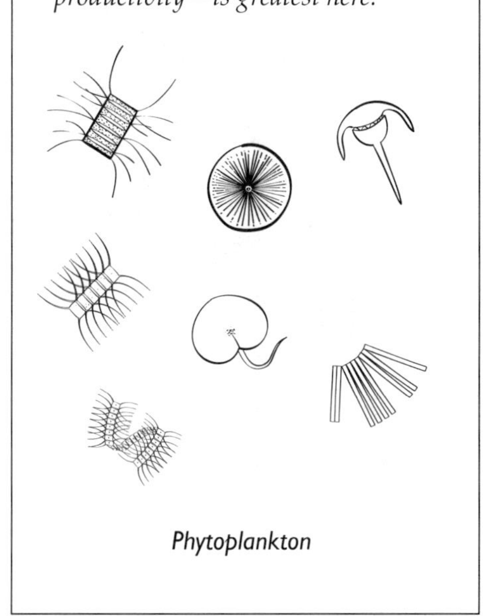

Phytoplankton

Wrap-Up

1. Have students hold up the two tiniest slices they saved. One sliver represents our **drinkable water,** the resource necessary for all life on land; the other represents all the **upwelling zones,** among the most productive places on Earth. Compare these two slivers to the whole apple. Ask, "What does the comparison of these slices to the whole apple tell you about these resources?" [These two minuscule pieces of our planet support nearly all of its life. They also represent the parts of the land and ocean that humans come in contact with, use, and affect the most. These two critical resources need to be protected to ensure a healthy future for our planet.]

2. Invite students to eat their apples as they begin to clean up. Let them know that since there's so little farmable land on which to grow all our food, you don't want to just throw away the uneaten apples. Ask, "What are other ways we could reuse these apples?" [Feed to animals, place in compost, etc.]

3. Assign students to collect each of the items they've been using (ocean pictures, markers, knives, apple pieces, paper towels). Give any uneaten apple pieces to students who offer to recycle them.

This is how your circle graph will look after all the sections have been added.

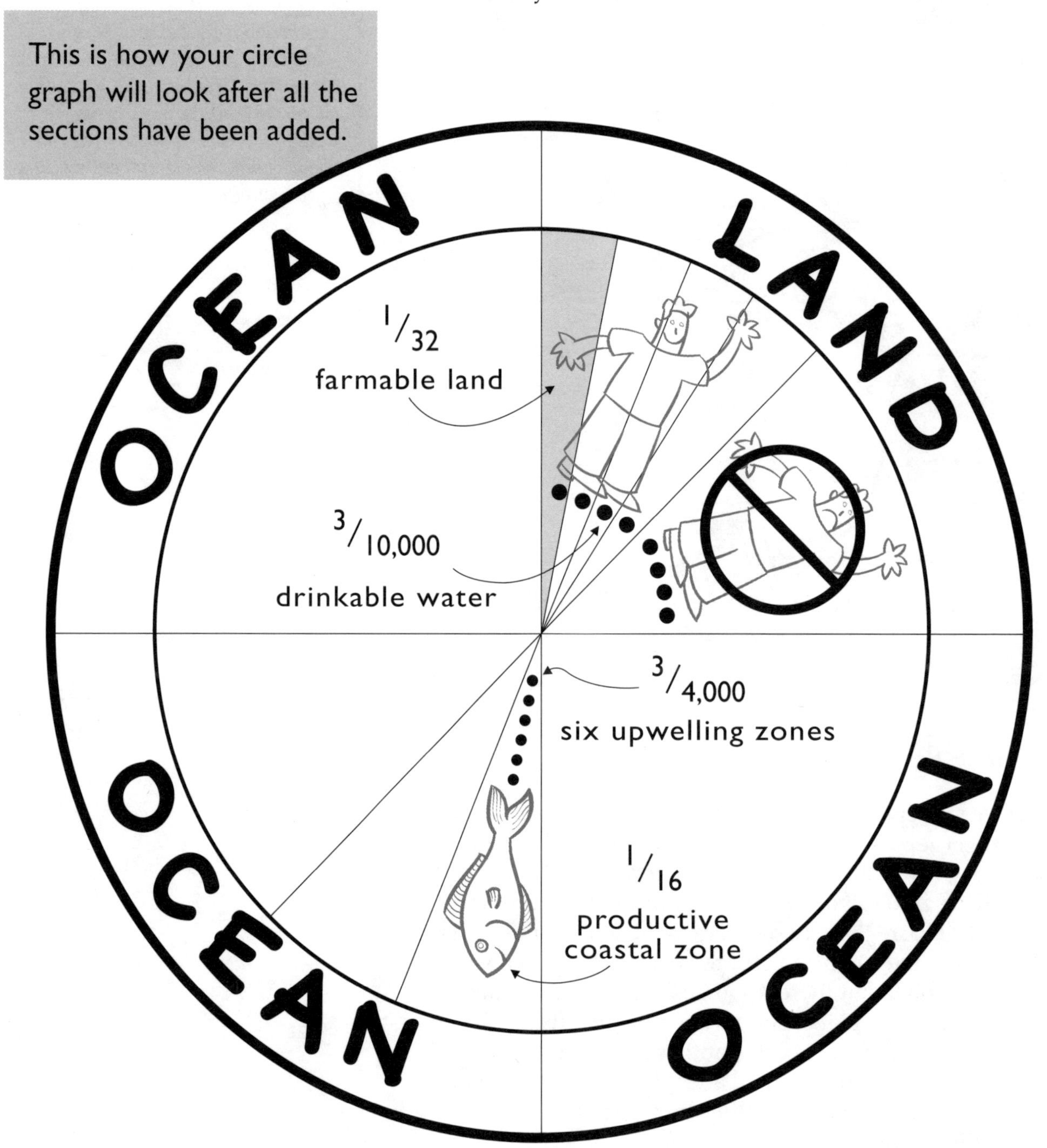

Think, Pair, Share: What Did We Learn?

Participating in a Think, Pair, Share activity allows students "think time" to formulate their own ideas and jot them down as notes. This gives all students, even those reticent to answer questions, the opportunity to organize their thoughts before hearing the answer from a classmate. This activity structure also allows the students to try out their ideas in progressively "less safe" settings: first with one partner, in the "safest" setting; then with a small group; and finally with the class.

1. Tell students that they've learned a *lot* about our planet, and it might be easy to forget some of the details if we don't review them. Ask students to take a moment to **"Think"** of the things that most surprised or interested them about "Planet Ocean" and have them write down as many as they can (at least three).

2. Next, have students **"Pair"** with a partner and **"Share"** their notes with each other.

3. Have each pair create a list of five new things they learned about the land, the ocean, or the planet as a whole.

4. Initiate a class brainstorm for sharing the lists. As students share their ideas with the class, record their thoughts on the class Brainstorm Chart for Session 2 (or overhead transparency); add new sheets as necessary. As you're writing, pairs should add new items from other pairs' lists to their own. Compare the new chart with the original class Brainstorm Chart from Session 1.

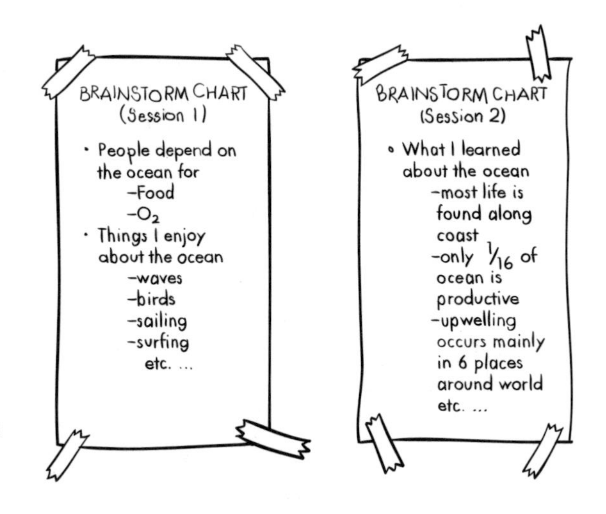

5. In closing this session, hold up the Key Concept for this activity and have one or more students read it aloud:

- **Most of our planet is covered by ocean, but only a small fraction of the ocean supports large concentrations of life.**

Briefly discuss how this statement reviews the important ideas from today's activities. Post the concept on the wall for students to revisit during the rest of the unit.

A Brief Introduction to Upwelling

(Can also be used as a special "Going Further" presentation.)

Upwelling is a critically important, seldom taught, and difficult to understand concept. It's worth spending a little extra time explaining, in order to help students understand it.

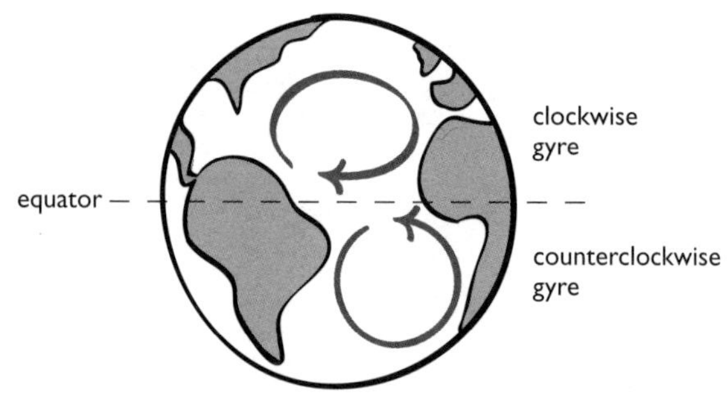

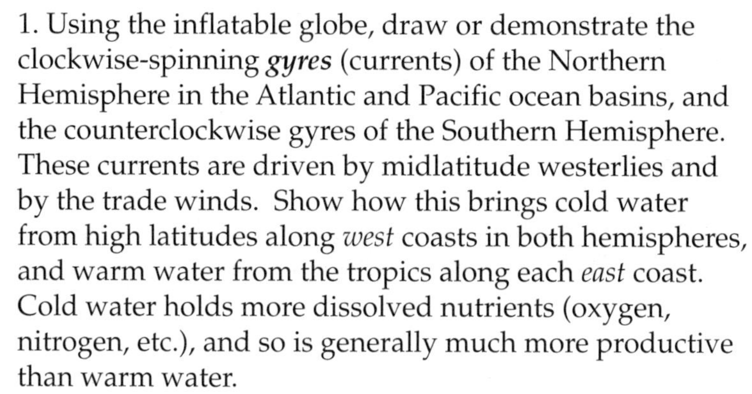

1. Using the inflatable globe, draw or demonstrate the clockwise-spinning ***gyres*** (currents) of the Northern Hemisphere in the Atlantic and Pacific ocean basins, and the counterclockwise gyres of the Southern Hemisphere. These currents are driven by midlatitude westerlies and by the trade winds. Show how this brings cold water from high latitudes along *west* coasts in both hemispheres, and warm water from the tropics along each *east* coast. Cold water holds more dissolved nutrients (oxygen, nitrogen, etc.), and so is generally much more productive than warm water.

People are often confused by the use of east and west in this context. "East coast" means the eastern side of the land mass, on the western side of the ocean.

Note: Wind terminology ("north-westerly," "easterly," "southerly," etc.) is based on the direction from which *the winds are coming. We learn a lot about what weather to expect by where winds come from. (The west coast of North America is very long; there are many wind regimes at work, not just northwesterlies.)*

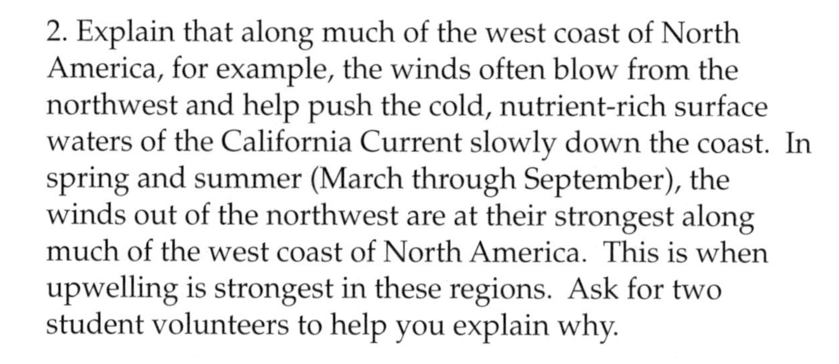

2. Explain that along much of the west coast of North America, for example, the winds often blow from the northwest and help push the cold, nutrient-rich surface waters of the California Current slowly down the coast. In spring and summer (March through September), the winds out of the northwest are at their strongest along much of the west coast of North America. This is when upwelling is strongest in these regions. Ask for two student volunteers to help you explain why.

3. One student can be "Atlas," and will hold up the globe and slowly spin it on its axis from west to east. The other can be "Aeolus" (Ruler of the Winds). Give Aeolus a marker, and have her "blow" (draw) a surface current down the west coast of North America as Atlas slowly spins the world from **west to east.** Aeolus can only move her pen from **north to south.** Try it a few times.

4. As the wind blows north to south and the Earth spins west to east, the marker should trace a "current" moving to the west (offshore) at nearly a right angle to the wind. This is similar to what happens on the ocean. The harder the wind blows, the more surface water appears to be moving offshore (actually, the land is spinning away from the water!). As the surface water moves offshore, it's replaced by very cold water "welling up" from deep down—an upward current.

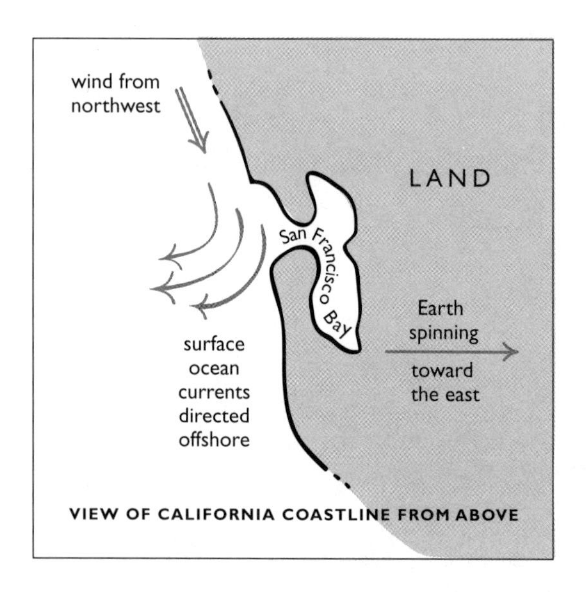

5. All year long, plantlike organisms and animals die and decompose and slowly sink out of the sunlit photic zone, toward the bottom. (Scientists call this the "rain of bodies.") Upwelling water carries a lot of this debris ***(detritus)*** from dead, decomposed plantlike organisms and animals. It acts as a super-rich fertilizer just at the time of year when day length is peaking. The entire food web blooms in dramatic fashion. First the water turns pea-soup green—full of microscopic plantlike organisms, or phytoplankton. Microscopic animals, or zooplankton, graze on the phytoplankton and multiply. Fish larvae (also considered zooplankton) hatch out and begin eating the smaller zooplankton. Seabirds migrate by the hundreds of thousands into the area to lay their eggs and feed on the fish larvae. Marine mammals—whales, dolphins, porpoises, and seals—also migrate into the area in huge numbers, to feed on the fish and zooplankton.

The exciting MARE guide Ocean Currents *includes a vivid demonstration of the effects of wind blowing over the surface of the ocean. As students blow across water in a pan, they create circulating currents, or* ***gyres****—reproducing what happens on a vast scale when strong winds sweep across the surface of the ocean.*

6. Thanks to upwelling, the Central California coast is one of the most productive places on Earth. Alarmingly, California has already lost 95 percent of its coastal wetlands. In an effort to save what's left, many stretches of coastline have been protected as National Marine Sanctuaries. From approximately Bodega Bay south to San Simeon, the coast is protected from certain activities (such as drilling for oil) by three contiguous sanctuaries: Cordell Banks, Gulf of the Farallones, and Monterey Bay. A fourth sanctuary protects the Channel Islands. Draw these areas on the globe and on your poster/map (if you have one) of the west coast of North America.

Extra notes on conserving our sensitive marine regions:

7. Many people support protecting unique ocean regions. But while nearly 60 percent of people in the U.S. think the ocean is in trouble, most don't understand how "boundaries" can be set in the ocean—and people assume more areas are protected than actually are. As we've learned, while the ocean makes up three-fourths of the Earth's surface, less than one percent of the marine environment is officially protected. In 1977 the National Marine Sanctuary program was established to protect marine sites (much as the National Park Service protects important places on land). Sanctuaries now protect portions of the marine environments off California, Washington, Massachusetts, Georgia, and North Carolina; in the Gulf of Mexico; and in the Florida Keys. Sanctuaries protect the humpback whale in Hawaii and a coral reef in American Samoa. One of the Great Lakes, Huron, has created a sanctuary that protects

historic shipwrecks. Draw these locations on your globe or world map. Ask students how *they* would decide where to create marine sanctuaries.

8. Although the main purpose of sanctuaries is to protect special environments, a strong secondary purpose is to support multiple uses, including fishing. But taking resources in sanctuaries is a controversial issue. "No-take" marine reserves have been proposed for some coastlines in order to keep them relatively untouched, help protect the ocean's biodiversity, and allow areas to recover from over-exploitation. Places such as Samoa, Tanzania, and Vietnam have set up Marine Protected Areas (MPAs) to protect biodiversity and sustain the human communities that depend on local marine resources. In order for MPAs to work, the communities, the tourism industry, and the fishing industry must strongly support the idea. These are the people whose livelihoods depend on the health of the marine environment. Ask students to research MPAs around the world and label them on your globe or world map.

Session 3: Creative Writing

To help students organize what they've learned about the ocean in a fun and creative way, select one or more of the following activities to do with your class. These activities provide opportunities for students to use written language in meaningful ways—and for you, the teacher, to assess their science knowledge and writing skills. These informal formats for writing are especially appropriate for English-language learners who may be intimidated by more standard writing assignments.

Mini-books should be written in whatever language students are most comfortable using, and the focus should be on content and creativity, not grammar and spelling. Mini-books can be introduced and begun as a whole-group activity and completed during subsequent work sessions or whenever students have extra time. Consider letting students finish during your scheduled writing time. Mini-books are ideal assessment tools; collect them, develop an evaluation or scoring system, and include them in student portfolios.

A. Mini-Book on the Ocean

1. Tell students that they get to become authors of a book about "Planet Ocean" by making a mini-book with illustrations and text. First they need to make the blank book. Pass out scissors, fine-point markers, and either 10" x 14" or 11" x 17" paper, and lead them through the directions on the appropriate mini-book instruction (pages 38–40); there are instructions for simple and more complex versions, depending on your preference.

2. With the book folded shut and only the "cover" showing, have each student author create a title for his book on the cover, illustrate it, and add his name. Encourage students to write this book about what they've learned by creating text and illustrations on alternate pages. The written and pictorial descriptions should relate to one another. They can label one section or chapter "The Planet," one "The Land," and one "The Ocean." Have students be sure to include information related to the Key Concept for this activity and include in each chapter at least four new things they learned.

3. Provide students ample time to write and draw pictures appropriate for each chapter.

4. Allow time for students to share their books with others and check one another for accuracy and understanding.

5. Create a mini-book library for students to use in future research projects on the ocean.

B. Posters, Journal Writing, Comic Books, and Travel Brochures

1. Create a list of topics, scenarios, and assignments related to the Apples and Ocean activity for students to use as the core of their writing or drawings in this session. Even better—have students come up with their own ideas! Some possibilities include:

- Create an educational comic book to tell people what upwelling is and why it's so important to the planet.
- Create a poster communicating the small fractional portion of our planet that supports nearly all life.
- Create a poster or brochure encouraging people to protect our fresh water and/or our upwelling areas.
- Write or draw a booklet that responds to these questions: What are some ways we depend on the small amount of drinkable water available to us? How many different ways do we use this water? Where and in what ways do we waste it?

If you introduced the concept of the photic zone, here are some additional topics:

- Design a travel brochure describing a trip to the photic zone and the zone below. Incorporate these questions: Why don't many animals or plantlike organisms live below the photic zone? What do you think nourishes them if they do live there?
- Write or draw a booklet that responds to these questions: What are some ways we depend on the shallow photic zone that lies in a narrow band along the coasts of the ocean? How do we use this zone?

2. Have students share what they create through poster sessions, or informally through the class library.

Going Further

1. Revisit your lists of questions from the Brainstorming Charts (and Question Chart, if you made one). If there are still unanswered questions or unresolved misconceptions, break students up into groups and assign a question to each group as a research project.

2. Have students work in cooperative groups to complete one of the following activities. Have each group present the results of its work to the entire class.

- List and discuss all the ways you can conserve drinkable water at school and at home.
- Look at a rainfall map of the world in an atlas, and at a regular map or globe showing rivers and lakes. Taken together, what regions have the most drinkable water? The least? How does this affect people's lifestyles, or where they live?
- Research the upwelling phenomenon. What causes it? Where does it occur? What animals are associated with upwelling areas? What happens during an *El Niño* year, when weather disruption includes a rise in sea-surface temperature and a drastic decline in ocean productivity?
- Research and discuss several cultural myths and stories about the open ocean. What do different cultures believe about the ocean? What do these say about the relationship between their culture and the ocean?
- Write to a National Marine Sanctuary office (see "Resources") to obtain information about the sanctuary system. Where are all the marine sanctuaries? When and why were they designated? What special plantlike and animal populations occur in each?

3. Have students record their circle graphs from Session 2 on grid paper and label all the parts. Be sure they add a title and legend that explain what the graph represents. As an additional assignment, each student can write a narrative description of what's represented on her circle graph.

Mini-book (11" x 17")

1. Fold the sheet in half crosswise.

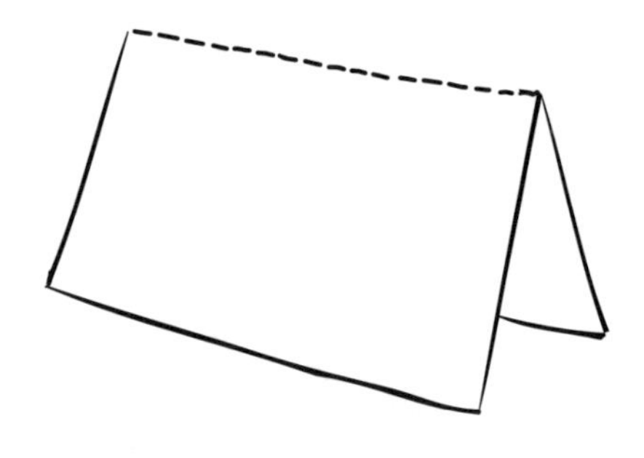

2. Fold up ends separately to form a "W" shape.

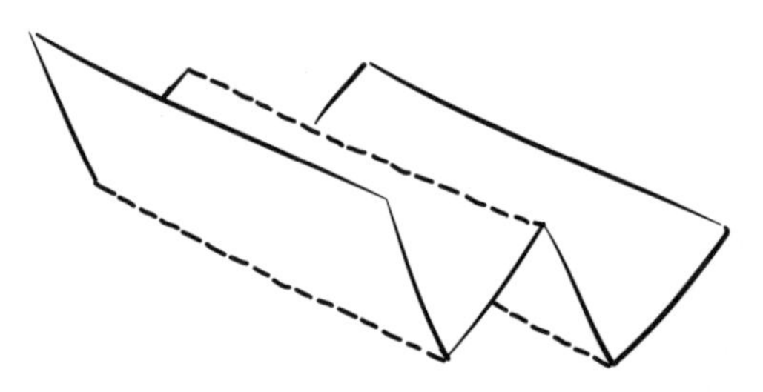

3. Fold the paper in half again to form a small rectangle. Then unfold this last fold, and fold it again back the opposite way, making good, hard creases on each side.

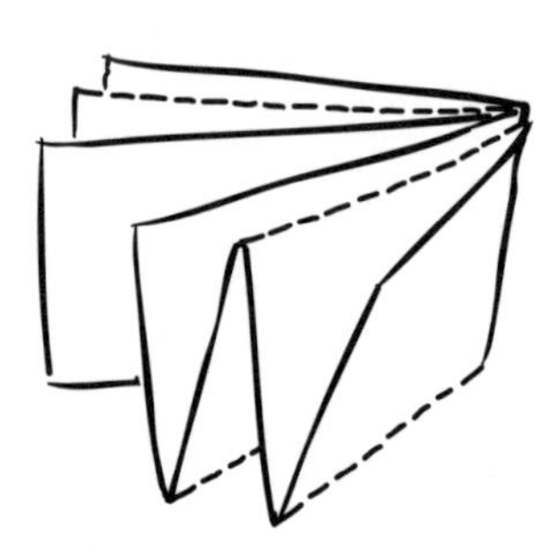

4. Unfold back to step #1, where the sheet is only folded in half.

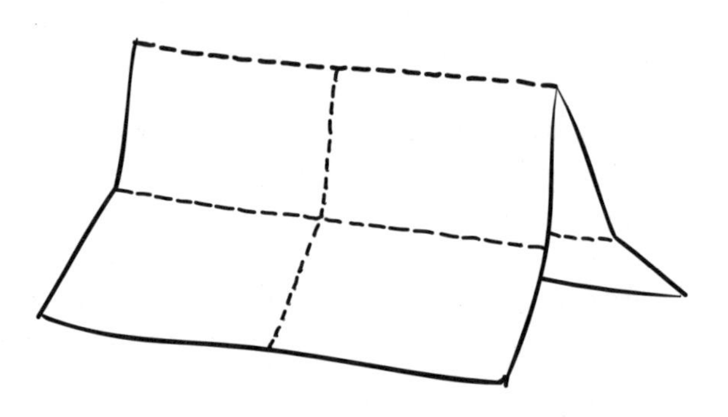

5. Face the folded edge closest to you and cut along the middle fold through both sides to the center as seen in the diagram.

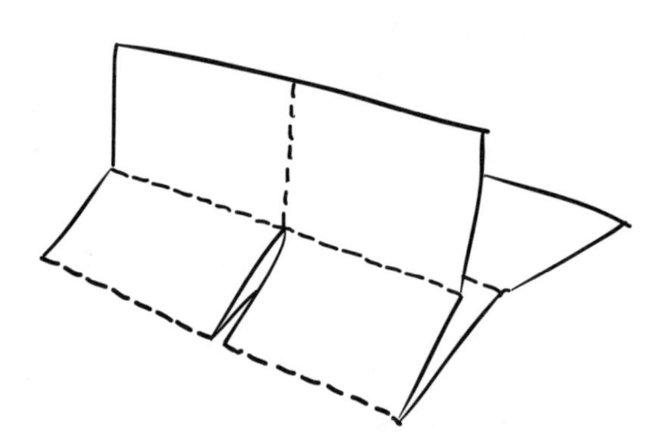

6. Unfold the sheet entirely.

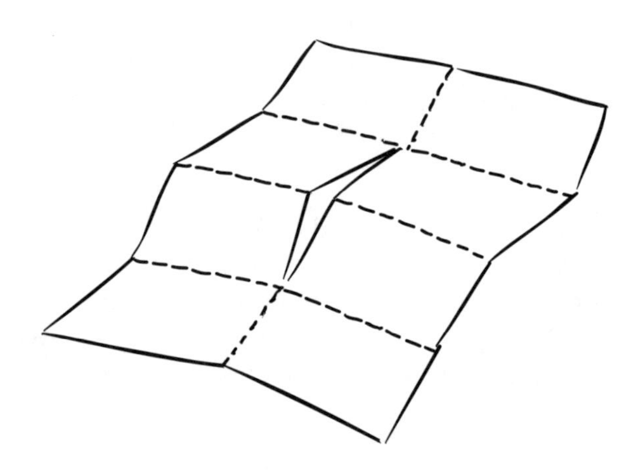

Mini-book (11" x 17") continued

7. Refold the sheet in half, this time lengthwise.

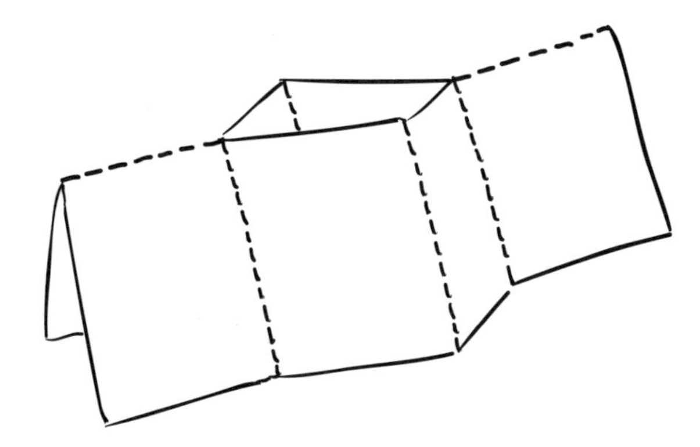

8. Grab the two outside panels and push inward. The part you cut with the scissors should open up and form a diamond.

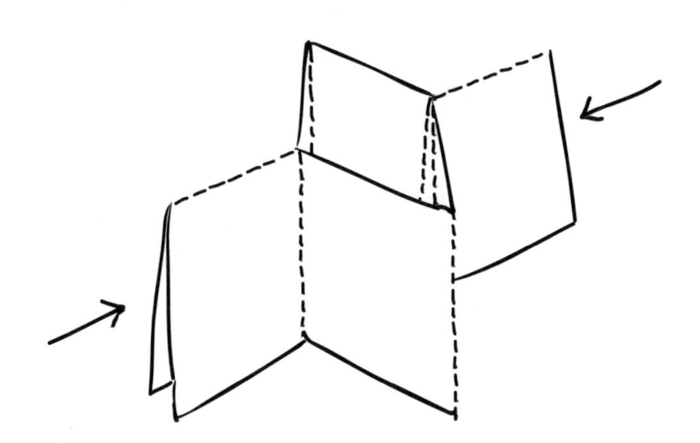

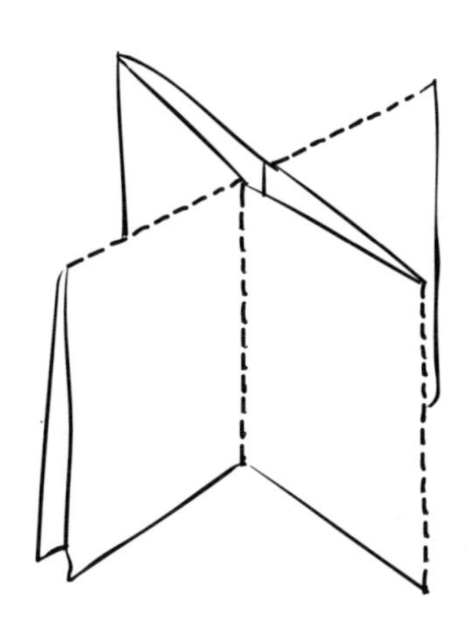

9. Finally, fold all the pages together to form a small book. Make good, hard creases on all sides.

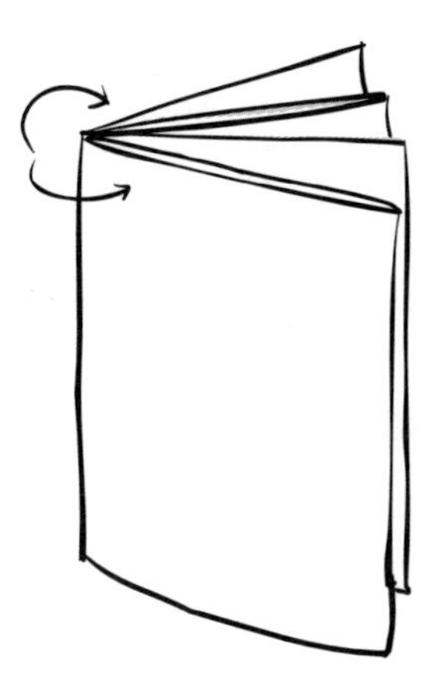

Mini-book (10" x 14")

Fold a 10" x 14" piece of plain paper in half lengthwise and then into thirds. Open it up so it is only folded in half lengthwise—with the fold on top. On the top half only, use scissors to cut along the two small folds to form three flaps that open vertically. Then fold the right third to the center, and the left third on top of that. With the book folded shut and only the "cover" showing, have students write the title of their mini-book on the cover and illustrate it. Then have them open the cover (from right to left) and write their name as the author. Then turn the title page (from left to right) and label each of the three chapters or sections. As they flip up each of the three chapters they can use one panel inside to draw a picture and the other to write about what they've learned.

1. Fold in half lengthwise.

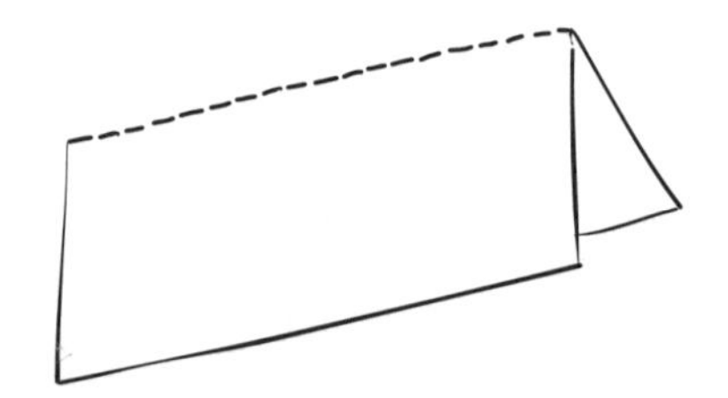

2. Then fold into thirds.

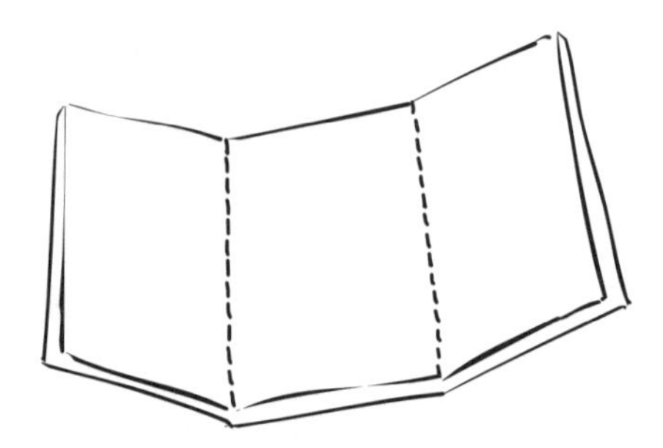

3. Cut only the top half into three sections.

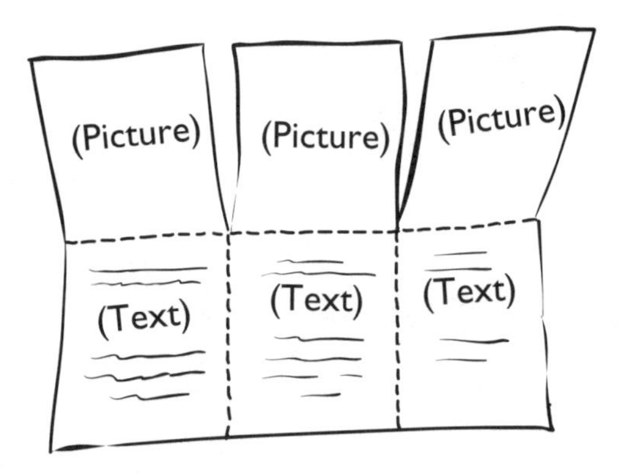

4. It will now look like this.

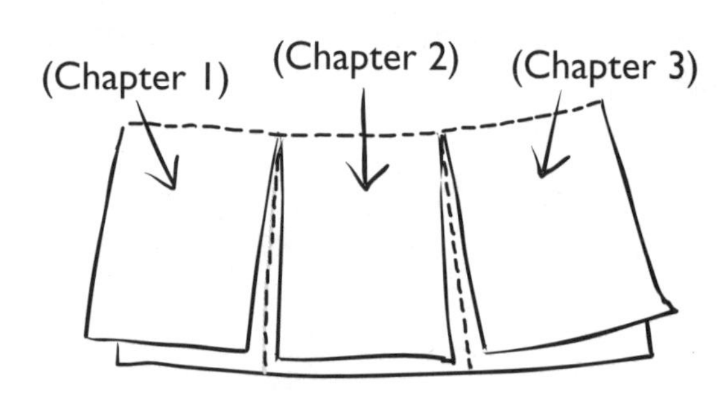

5. Fold the right third to the center and the left third on top of that.

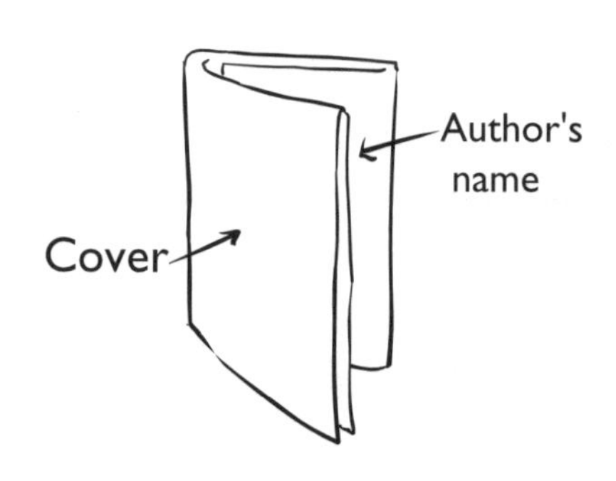

APPLE = PLANET
LAND
1/4 land
1/8 uninhabitable
1/8 habitable
1/16
1/16
1/32
1/32 farmable land...

1. ARM
Count how many? 10
6. COLLAR
12. FIN
5. SIPHON
4. EYE
3. BEAK

Activity 2: Squids—Outside and Inside

Overview

Our vast ocean covers three-fourths of the planet, yet only a fraction of the ocean (1/1,000) supports the large concentrations of life that people depend on for food and livelihoods. In this activity, students will observe in depth (and inside!) one ecologically and economically important species—the squid.

Our knowledge of squids has tentacles reaching far back in history. Sailors of old described them in their yarns and legends, most often as monsters of the sea, or "kraken." Squids are related to the octopus, cuttlefish, clam, and snail—all classified as phylum Mollusca. Squids and other members of the class Cephalopoda (meaning "head-foot") have the most highly developed nervous systems of any invertebrates, and the most complex behavior. Their sense organs, brain, and excellent swimming ability even allow them to compete with vertebrates for food and territory. Squids are called ***pelagic*** *creatures because they're adapted to a completely oceanic way of life. They have gills, a streamlined body, a reduced shell, and a funnel (siphon) for speedy propulsion through the water.*

Squids are found in every ocean in the world and are eaten by many different predators. They're also a very important food resource for people around the planet. This edibility in turn makes them excellent dissection animals; it's a better use of our world's limited resources—and perhaps shows more honor toward the organism—to dissect something that can also be used in other ways afterwards.

In this activity, students work in pairs to dissect a squid. They investigate its ***adaptations:*** its structure and how all the parts function together to allow the squid to survive and thrive in its open-ocean environment. The squid is then honored as the students participate in a Calamari Festival, creating another use for the organism they dissected; in this way we attempt to "complete the cycle" and reinforce the concept of the importance of each organism to the web of life. In the last session, the class explores the issues surrounding the threatened squid fishery by role-playing and discussing the problem from different points of view. Students come up with possible solutions to an important global problem.

As emphasized in the National Science Education Standards, *students need to understand the intricacies of structure and function in living things. Dissecting squids offers a special opportunity to view and appreciate a unique animal's characteristics and adaptations up close.*

In Session 1, students get "**Into** the Activity" in small groups. Before and after watching video footage, they read and discuss several statements about squids. They record their knowledge and any questions they may have about squids on their own Anticipatory Charts. They share their prior knowledge with others; first with a partner, then in a foursome, and finally with the class. In Session 2, students move "**Through** the Activity" as they work in pairs to dissect and draw their own squids, as the teacher models each step and draws the parts on the class chart. The dissected squids are then cleaned by the students, cooked, and served in a squid feast. In Session 3, students participate in debriefing games, including Squid Pictionary and Squid Jeopardy, to reinforce concepts and check for understanding. Students discover that looking closely at an animal like the squid can tell us a lot about the adaptations needed to survive and thrive as a pelagic (open-ocean) creature.

In Session 4, students look more closely at the socioeconomic side of the squid fishery (commercial fishing operation), as cooperative student groups represent the viewpoints of different interest groups at a Squid Fishery Symposium. In this session students discover that many people depend on squids for food or to make a living, and that more discussion among these people will help create solutions to the problem of diminishing squid populations. The "Going Further" section allows the teacher to take students "**Beyond** the Activity" with optional extension activities, and a take-home handout suggests additional activities for the family.

What You Need

For the teacher:

- ❑ 1 squid for your practice dissection (see "Getting Ready," page 46)
- ❑ 3 squid anatomy charts: 1 "External," 1 "Internal," and 1 "What We Learned" (see "Getting Ready," page 47)
- ❑ *(optional)* 1 copy of the Squid Dissection Summary Outline for Teacher's Notes (master on pages 80–81)

You may prefer to use an overhead projector to draw the squid anatomy or even make an overhead from the unlabeled Internal and External Squid Diagrams in the guide. Any method will work as long as you're comfortable with it. We suggest, however, that not only is it good modeling of instruction and biological illustration to draw the charts yourself in front of the class, but the large multicolored charts are very appealing, and can be left up on the wall to serve as a reminder of the impressive amount of knowledge gained by the students. They can also be referred to later on, during the debriefing.

For the class:

- ❑ a videocassette recorder (VCR)
- ❑ a viewing monitor
- ❑ Monterey Bay Aquarium Video Collection: "Seasons of the Squid," or other video containing squid footage (see "Resources," page 163)
- ❑ 10 sheets of chart paper (approximately 27" x 34")
- ❑ 1 set of colored markers (wide-tipped, water-based)
- ❑ 1 small jar with lid (like a baby-food jar) half-filled with water
- ❑ 1 Squid Statements Anticipatory Chart (example on page 82)
- ❑ 1 Squid Interest-Group Chart (example on page 48)
- ❑ 1 About Squids Chart (example on page 46)
- ❑ a roll of masking tape
- ❑ paper towels
- ❑ *(optional)* 1 microscope
- ❑ *(optional)* 1 small dish (such as a Petri dish) to use with the microscope
- ❑ *(optional)* a laminator for the class Squid Statements

Anticipatory Chart and "Internal Squid" Chart (see "Getting Ready") and Squid Interest-Group Profiles (see below)

If you plan to cook the squids:

- ❑ enough packages of tempura batter mix (the just-add-water kind) for the class (see page 71)
- ❑ 1 "Fry Daddy" or other deep-fat fryer
- ❑ cooking oil for frying
- ❑ 1 medium bowl
- ❑ 1 spoon to mix the batter
- ❑ 1 knife to cut the squid
- ❑ 1 cutting board
- ❑ *(optional)* lemon for garnish

For each of six small groups of students:

- ❑ 1 Squid Statements Anticipatory Chart student sheet (master on page 82)
- ❑ 1 Squid Interest-Group Profile student sheet (master on pages 85–86) cut into individual profiles
- ❑ 1 sheet of lined 8 ½" x 11" paper

For each pair of students (plus one extra for the teacher):

- ❑ 1 squid (see "Getting Ready")
- ❑ 1 pair of scissors to cut the squid (see "Getting Ready")
- ❑ 2 toothpicks
- ❑ 2 **stacked** full-sized, round, white, undivided, sturdy paper plates (Chinet® or similar); the bottom plate will be used to put the cleaned squid on
- ❑ 2 or more paper towels
- ❑ 1 hand lens
- ❑ 1 ruler
- ❑ 2 sheets of blank 8 ½" x 11" paper

For each student:

- ❑ 1 sheet of lined 8 ½" x 11" paper
- ❑ 1 unlabeled External Squid Diagram student sheet (master on page 83), if you play Squid Jeopardy (see page 72)
- ❑ 1 unlabeled Internal Squid Diagram student sheet (master on page 84), if you play Squid Jeopardy

Getting Ready

1. Squids are available in grocery stores and Asian food markets, frozen in five-pound boxes, for about $1–$1.50 per pound. There are typically 30–50 squids per box; enough to cook some while students are dissecting the rest. Squids are also available fresh, in fish markets. (Frozen squids are the number-one choice, but if only fresh are available, choose the largest and freshest squids—and make sure they haven't been cleaned!) Whichever you buy, keep them in the freezer until the morning you plan to use them. They'll thaw under cold, running water. If you don't find fresh or frozen squids, your supermarket should be able to order them frozen from distributors who work directly with suppliers on the West or East Coast. **Do not use preserved squids. The internal structures aren't as easily dissected, and you *cannot* eat squids preserved with chemicals.**

If students' hands are clean, tools have been washed, and the dissected squids have been rinsed, the squid should be perfectly safe to eat after the dissection. Teachers have reported that the Calamari Festival is well worth the effort, and that it emphasizes the need to respect and reuse the squids.

2. If you'll be cooking the squids after the dissection, (which we strongly suggest!), students must wash their hands with soap and water before the dissection and the scissors must be washed and disinfected before use. (It works well to take the scissors home and put them in the dishwasher before—and after!—the dissection.)

Squid Statements Anticipatory Chart

Statements About Squids	YES	NO	Don't Know
Squids don't change their color			
Squids are usually solitary			
Male and female squids look the same on the outside			
Squids have 10 arms that all look alike			
Squids use fins for locomotion			
Squids chase and catch fish			
Squids lay single, individual eggs			
Young squids look just like their parents, except smaller			
Squids jet through the water with their arms trailing behind			

3. Using the Squid Statements Anticipatory Chart student sheet on page 82 as a guide, draw a class chart on chart paper using colored markers. *Note: The statements shown on this chart refer to specific footage shown on the Monterey Bay Aquarium video; if you're using a different video, you may need to tailor the statements to what's portrayed in the footage.* You might want to laminate the chart so you can reuse it at a later date. If you decide to laminate, be sure to use only washable markers when recording data. (Mr. Sketch® wide-tipped markers work well on laminated materials.)

ABOUT SQUIDS

What we think we know	What we'd like to find out	What we know to be true

4. Make a class About Squids Chart: Using chart paper (turned sideways) and colored markers, write "About Squids" at the top of the paper. Draw two lines down the chart paper to divide it into three columns. At the top of the first column, write "What we think we know." At the top of the second column, write "What we'd like to find out." At the top of the third column, write "What we know to be true."

5. **Read through the entire dissection activity and then dissect one squid for yourself as practice before trying it out with your students.** You'll probably be amazed at

how easy and straightforward a squid dissection can be. **Determine if you want to present the dissection in its entirety, or if you'd prefer to leave out some of the vocabulary and explanations** (perhaps especially as concerns the reproductive system). For easy reference during the dissection, you may want to use the Squid Dissection Summary Outline for Teacher's Notes (pages 80–81) filled in with your own guiding language.

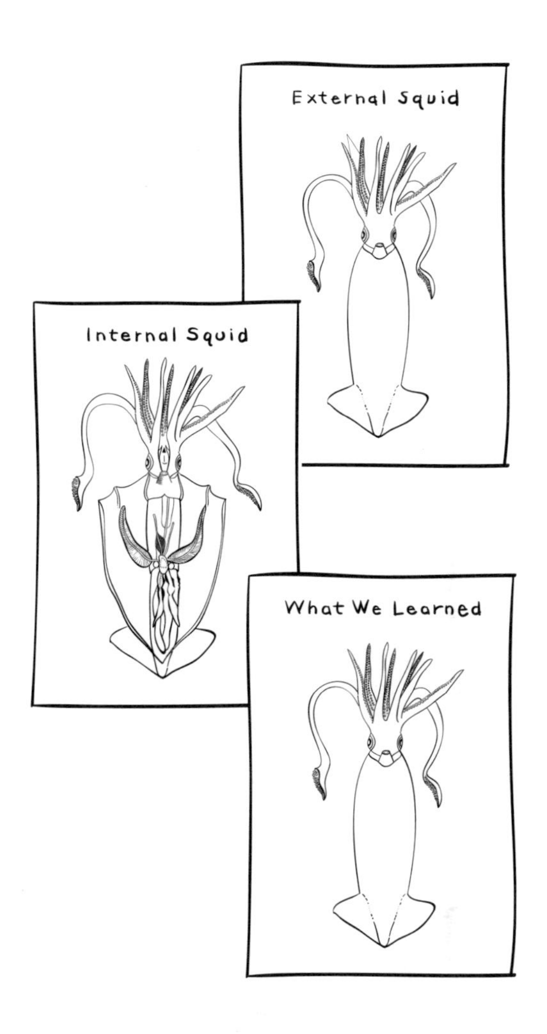

6. Draw the following squid charts on chart paper using the colored markers. Draw the charts BIG, so you can add features to them as you lead the dissection. **Refer to the labeled External and Internal Squid Diagrams on pages 78–79 to locate all the parts as you're preparing to teach this activity.**

- squid chart #1: draw an outline of the squid and title the chart "External Squid."
- squid chart #2: draw an outline of the cut-open squid and title the chart "Internal Squid." You may want to use very light pencil to draw in the internal organs on the chart and then laminate it—it's then a simple matter to trace over the pencil with markers as the dissection proceeds, so students can see the organs.
- squid chart #3: draw an outline of the squid and title the chart "What We Learned." This chart will be used for review.

7. Duplicate the Squid Statements Anticipatory Chart (master on page 82) for each small group.

8. If you've decided to play Squid Jeopardy (see page 72), duplicate one unlabeled "Internal Squid" and one unlabeled "External Squid" Diagram (masters on pages 83–84) for each student.

You might use this as an opportunity to get more parents involved in your classroom. Ask if one or more of them can come the day of the dissection and cook the squids using a favorite recipe. There are many recipes for cooking squid; use your own or deep-fat fry it as in this activity (see page 71). We use the just-add-water tempura batter and kids seem to love it. If you're unable to get helpers to cook the squids, you might use this as a cooking lesson for the students, cook it over recess or lunch, or take it home with you to cook and bring back to class the next day.

9. Determine when you'll cook the squids. It's ideal if an aide or parent volunteer can cook the extra (undissected) squids while the dissection is underway, so the squids will be ready to eat when the dissection is complete. The rest of the squids, dissected and cleaned by the students, can be cooked as the students watch, help, or go to recess.

10. Place two toothpicks, a hand lens, a ruler, and a pair of scissors on each top plate. Just before the activity, **place the squids on the plates so that the funnels and the lighter side of the bodies are facing up and both fins are**

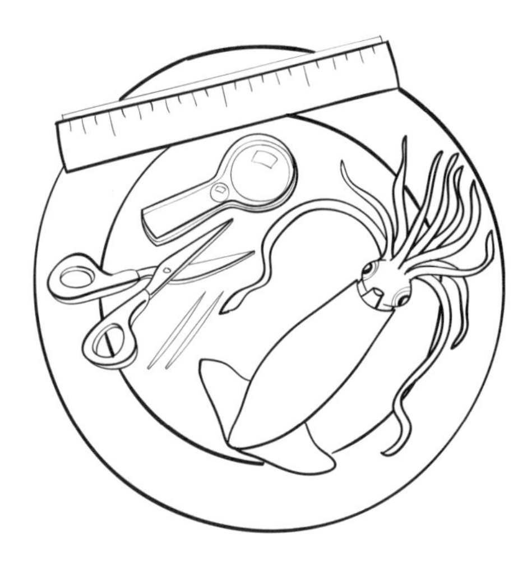

flat against the plate. If all the squids are oriented in the same way, it'll be easier for the students to match the location of and find specific parts on their squids as you show them on the class charts and use your own squid as a model.

11. Write out the Key Concepts for this activity in large, bold, letters on separate strips of chart paper and set aside.

- **Pelagic creatures are organisms living in the open ocean.**
- **Looking closely at an animal like the squid can tell us a lot about the adaptations needed to survive and thrive as a pelagic creature.**
- **Many people depend on squids for food or for their livelihood. More discussion among these people will help create solutions to the problem of diminishing squid populations.**

12. Tape together the short sides of two sheets of chart paper (turned sidewise) to make one long sheet. Use colored markers and write the title "Squid Interest-Group Chart" across the top of the paper in large, bold letters. Make seven columns and three rows, and label as follows:

SQUID INTEREST-GROUP CHART

	Squid Fisher	Consumer	Restaurant Owner	Sport Fisher	Biologist	Environmentalist
#1 How would we know if squids were over-fished?						
#2 How can we prevent squids from being over-fished?						

13. Make a copy of the two-page Squid Interest-Group Profile student sheet (master on pages 85–86) and cut apart the six different interest-group descriptions so each group is given **only the description of its own viewpoint** (squid fisher, consumer, biologist, etc.). Laminate, if you wish.

14. Decide if you'll invite members of the school (another class, the principal) and community (parents, conservation groups) to the Squid Fishery Symposium. You might have the class design a flyer advertising the event.

15. Gather the rest of the materials and have them ready for use as you present the activity.

Session 1: What about Squids? Anticipatory Charts

The word "fisher" is used throughout this guide, rather than fisherman, out of respect for all the women who fish for a living or for sport.

Making an Anticipatory Chart is an activity structure designed to help students recall information from past experiences and help them clarify what they want to learn from the activities that follow. In the Anticipatory Chart activity, students work together in small groups and have opportunities to share their ideas with the whole class.

Squid Statements

1. Review the significance of ***upwelling*** to the areas in the ocean that support abundant life. [Upwelling brings cold, nutrient-rich water to the surface to act as fertilizer for the phytoplankton living in the sunlit surface waters. These zones are usually very rich with life.] Ask the students, "What kinds of life from the ocean do people harvest from these areas?" "What is a fishery?" "What kinds of ocean food do you eat at home?" If no one mentions squids, tell the class that the squid fishery is an important food industry based in the productive coastal areas of the ocean—the "$^1/_{16}$" of the planet's ocean they looked at in Activity 1, Apples and Oceans.

Did You Know?
"Shoals" *of fish or other marine organisms are groups that have come together for social reasons—to mate or to feed, for instance. There's no pattern to the way they swim together.*

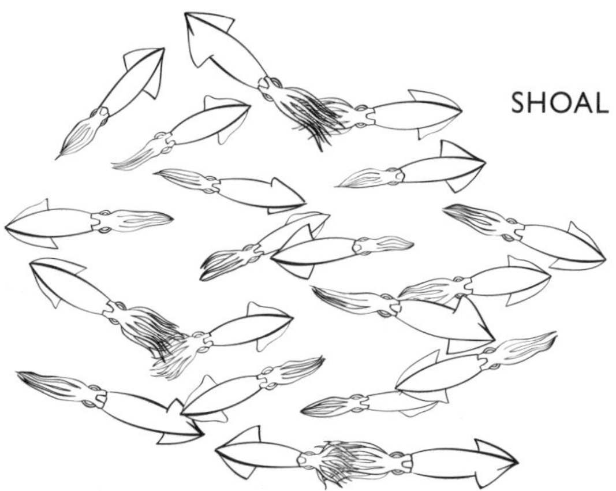

"Schools" *of fish or other marine organisms are shoals whose members 1) "orient" (point) in the same direction, 2) swim at the same speed, and 3) move parallel to each other. So "schooling," in the ocean, is really "synchronized swimming"—creating a pattern with your neighbors in order to fool or escape predators, etc. We often mistakenly use "school" when we mean "shoal"!*

2. Tell the class they'll be studying squids and the squid fishery, and say "Let's see what you already know about this animal!"

3. Divide the students into small groups for this activity. Distribute one copy of the Squid Statements Anticipatory Chart student sheet to each small group. Tell the students that as they discuss each of the statements, one person can act as the Recorder to mark down the number of students in their group who answer Yes, No, or Don't Know. **Stress that in science, discovering what we don't know is as important as what we *do* know—so no one should worry about her breadth of knowledge.**

Squid Statements Anticipatory Chart

Statements About Squids	*YES*	*NO*	*Don't Know*
Squids don't change their color			
Squids are usually solitary			
Male and female squids look the same on the outside			
Squids have 10 arms that all look alike			
Squids use fins for locomotion			
Squids chase and catch fish			
Squids lay single, individual eggs			
Young squids look just like their parents, except smaller			
Squids jet through the water with their arms trailing behind			

The vocabulary used in many videos is given in a too-much/too-fast mode, which causes many students to lose interest. Even more often, students perceive videos as an opportunity to relax and day-dream—much as television is used at home. We suggest changing this perception by using videos to take the students on a "virtual field trip." In order to do this you'll first need to turn the sound off and give the students directions in what you want them to look for. Students are more observant when they must look for action and hypothesize about its relevance, instead of just waiting to hear it described. This also gives the teacher the opportunity to direct the students' attention to specific topics, rather than just letting them go where the voiceover takes them. The teacher can have the students answer or discuss specific questions within their groups, sketch what they observe, and/or describe what they think is happening. You may find it helpful to use "white noise," such as an audiocassette of ocean sounds, to get the students over the hurdle of "listening" to the silence of the video.

4. As the students work, tape the class anticipatory chart up near the video monitor. Have the squid video in the VCR, ready to show. After each group has recorded its responses on its own group chart, turn on the video **(turn**

down the sound completely) and encourage the students to quietly discuss the statements within their groups in light of what they see on the tape.

5. At the end of the video segment, give each small group the opportunity to review the statements on its sheet and—if indicated—revise the number of students in the group who now answer Yes, No, or Don't Know.

6. Lead a class discussion and record the students' ideas on the class Squid Statements Anticipatory Chart. Ask the students which of the statements on the chart are left unresolved. Ask them how they might figure out if the statements are true or false. Tell them they'll be examining real squids, which will help answer many of their questions. Refer back to the chart throughout the activity as the students discover new information about squids.

Think, Pair, Share: "About Squids"

Having students record their ideas and questions on their own charts provides valuable practice organizing their thoughts in writing. It also provides you with an informal pre/post survey that can become part of student portfolios. If this type of recording is not appropriate for some of your students, then oral, small-group discussions, followed by whole-group sharing with you recording on the class chart, is a fine alternative.

1. Post the beginnings of the About Squids Chart. Distribute lined paper and have each student create her own chart. Have students first **"Think"** about each of the first two columns ("What we think we know" and "What we'd like to find out"), and then jot down and/or illustrate some of their own ideas.

2. Now have each student **"Pair"** up with another student, to discuss and compare ideas. Tell the students they can add to their lists after discussions with their partner if they like. Suggest that they might want to "star" the statements that appeared on both of their charts, to show that they had the same ideas.

3. Finally, have each pair **"Share"** its answers with another pair of students. Tell the students they can again add to their lists after discussion with the new pair of students, and can star ideas that appeared on both of the pairs' charts.

4. Lead a class discussion and record the group's responses on the class chart under "What we think we know" and "What we'd like to find out." Tell the students that as questions are answered, the third column, "What we know to be true," will be completed. Suggest that the students add to their own charts during the class discussion.

Session 2: Hands-On Squid

Special Notes for the Teacher

The squid dissection is one of the high points of this unit. In addition to the step-by-step procedures described below, there are several general considerations, based on our experience, that may prove helpful to keep in mind in presenting this exciting activity:

- It's best not to make any student participate in the dissection if he is really concerned or "squeamish." You might want to make a group of three if that's the case, and give the student the opportunity to take part later as he feels ready.

- Sometimes a few students will ask to use gloves before touching the squid. While we suggest that you discuss this with them, we strongly believe that "raw-hands-on" is best. There are some good reasons *not* to use gloves—including the simple fact that they're unnecessary, and very clumsy. Fresh squids do not smell at all fishy and are not the least bit slimy. This will probably surprise most students, since they may be familiar with the smell of fish and assume squids will be similar. Remind them that these were caught for human consumption, and if they were preparing them for dinner at home, they probably wouldn't use gloves.

- There are good reasons to have students work in pairs rather than individually, even though squids are not expensive dissection organisms. Students working together can help each other find all the parts, and they can take turns drawing and locating anatomy. It also helps reinforce the idea of respect; only the number of squids that will actually be cooked are dissected. We'll learn in another session that squids (and many other ocean food sources) may be overharvested.

- Working in teams also gives students the opportunity to use authentic language in meaningful ways in a less stressful environment. And it reduces your work; they can ask their partners to help them find squid parts, which avoids having 25 or 30 students each asking for your help. Another way to disperse the information quickly is to show a few pairs where the parts are located and then have them fan out through the class. (Show a different pair of students the parts each time, so everyone has an opportunity to share her newfound knowledge.)

Prelude to the Dissection

1. Discuss with the class the sanctity of life, referring back to the video and the beauty and grace of the squid. Let them know that today, they have the special opportunity to investigate the internal and external structures of real squids, up-close. Squids are a very important food resource for people—as well as for many other predators around the world. One or more species of squid can be found throughout most of the world's ocean, and they can be found in almost every grocery store around the world.

2. Tell students the squids they'll be dissecting are market squids, caught for human consumption. Although they're dead, they were once living, breathing organisms—the fastest and one of the smartest invertebrates in the ocean. We should show them the respect they deserve. Ask the students what it would look like to show respect to this creature during the dissection. [No poking or tearing, etc.; following directions.]

Even after numerous discussions about respect for once-living things, be prepared for some of your students to act inappropriately. We often tease and make fun of things we don't understand; leading the dissection with a spirit of "excitement for discovery" will help students become enthusiastic and focused explorers.

3. Ask the students what they think are good and bad reasons for doing a dissection. Tell them we're not simply going to throw away this once-living being after we're done with it; nor are we experimenting just for experiment's sake. We're going to study its ***adaptations.*** After the dissection we're going to have a squid feast, or Calamari Festival! It's better to dissect something that can be eaten afterwards (or used in some other way), to discourage a casual attitude toward taking an organism's life.

4. Stress that all your instructions need to be followed exactly, so that students can get the most out of the dissection and see all the parts. **No one is to use a tool until directed to do so,** and under no circumstances should they poke the squids. It shows disrespect—and also, the ink sac could be pierced, making the internal organs very difficult to distinguish. Remind the class that since these squids will be eaten later, students should work carefully and keep them clean.

The Dissection

1. After the students have washed their hands, pair them up and explain that they'll be working in twos so they can help each other find parts of the squid and take turns adding to their shared drawing. Distribute one sheet of blank paper per pair, for students to label "External

Squid." Pass out the stacked plates with the squids (correctly oriented) and the tools to each pair of students. Remind them to not touch their tools until told to do so. Place paper towels within reach of each student.

We urge you to use correct, authentic language (nidamental glands, ventral, dorsal, etc.) throughout the dissection, rather than simpler substitutes. Students "get" these terms, and should be encouraged to use them!

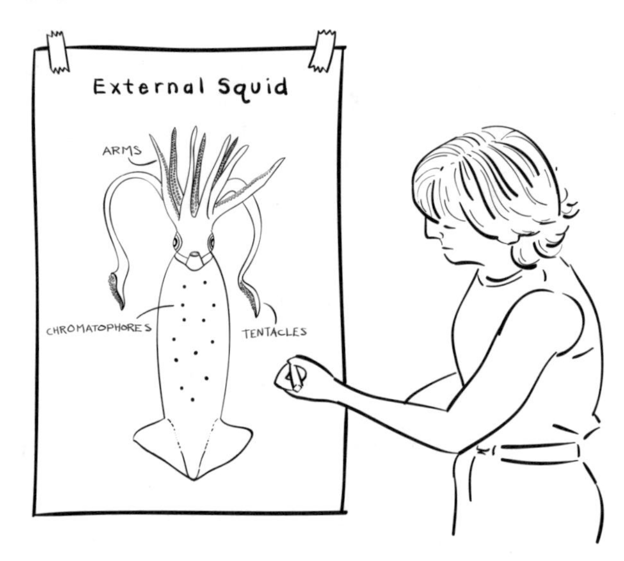

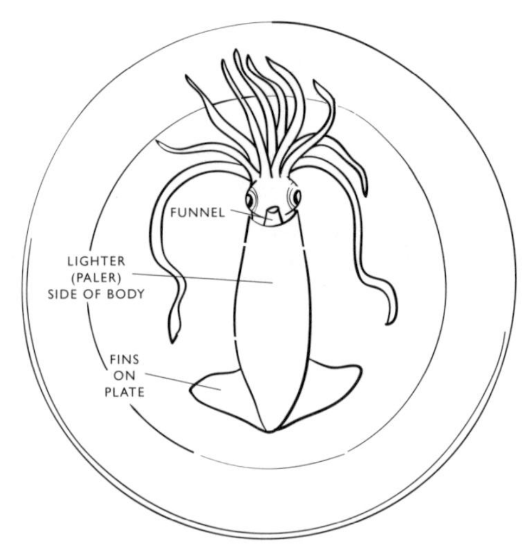

Squid correctly oriented—ventral side facing up

2. Tape squid chart #1, External Squid, and squid chart #2, Internal Squid, to the wall where everyone can see them. As each anatomical structure is introduced, draw the part on the corresponding chart and label it. Have students use their blank sheet of paper and draw and label their own chart along with you. This gives them good practice in taking notes and in biological illustration, and also serves as a check for understanding. Also, use your squid as a model to show the part and describe its function. **Be sure it's oriented correctly before you start.** Over the course of the dissection, walk around the room, helping students find the parts on their squids and checking for understanding.

Have students take turns as they draw the squids and participate in the dissection. You might suggest that one student draw the external anatomy, and the other the internal. Drawing their own squid helps students be more observant and helps them remember the anatomy.

External Anatomy: Arms, Head, Mouth, Eyes, and More

1. Have students gaze for a moment at their squids, looking at the relationship of the "head" to the rest of the body. Why do they think squids, octopuses, and their relatives got the name ***cephalopod,*** or "head-foot"? (*"Cephalo"* means head, *"pod"* means foot.)

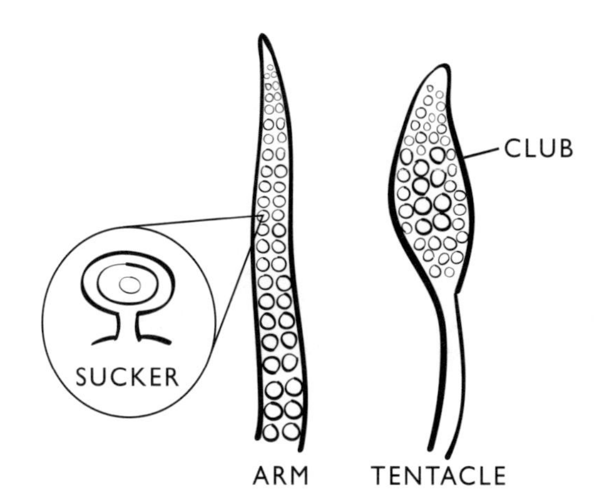

2. Begin the dissection. Ask students to count their squids' **arms** and determine if they're all the same—or, if not, how they differ from one another. (The differences are explained below.) This is a good time to use the hand lenses to get a close-up look at the **suckers.** Notice all the small teeth in a ring around each sucker and how the suckers are actually on short stalks. (Suggest that the students use the toothpicks to spread out the squid arms for finer control, or if they're not yet ready to touch the squids with their fingers.)

3. After students have counted the arms, ask them to describe what they saw. Tell them some "arms" are in fact called **tentacles.** Ask the students to hypothesize which ones they think should be called the tentacles and which the arms. [There are eight shorter arms lined with rows of suckers, and two longer, thinner tentacles with suckers located only on the tips, or **clubs.**]

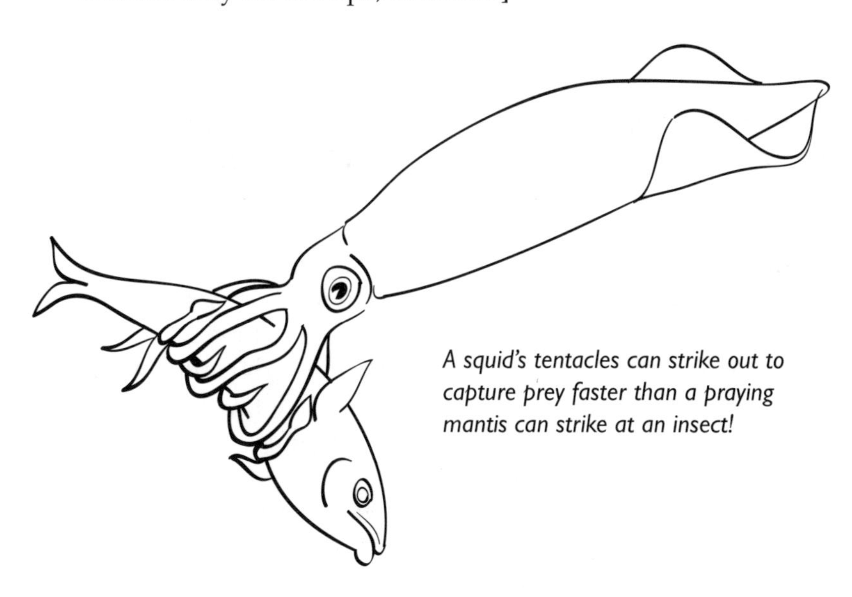

A squid's tentacles can strike out to capture prey faster than a praying mantis can strike at an insect!

4. Describe how squids capture their prey with the two long tentacles, then bring it into their "arm crown" of eight, suction-cup-covered arms, which hold the prey tightly as it struggles.

Squids consume up to 20% of their body weight in shrimp and fish per day.

Chitin *(pronounced ky-tin) is the same material that forms the hard exoskeleton (outer layer) of many insects and crustaceans—although it looks nothing like a crab shell. Chitin is similar to our fingernails.*

5. Have the students locate the squid's **beak**, made of chitin (a horn-like material), within the circle of the arms. The beak looks like a black speck, but on closer examination—using the toothpicks to push back the tissue surrounding it (called lips)—one can see that there are actually two halves to the beak, much like a parrot's beak. (You'll often see the plural, "beaks.") Before food is swallowed, the powerful beaks tear it into little pieces. (See more about this under Digestive System, on page 65.) The beaks are encased in a muscular ball called the **buccal mass.**

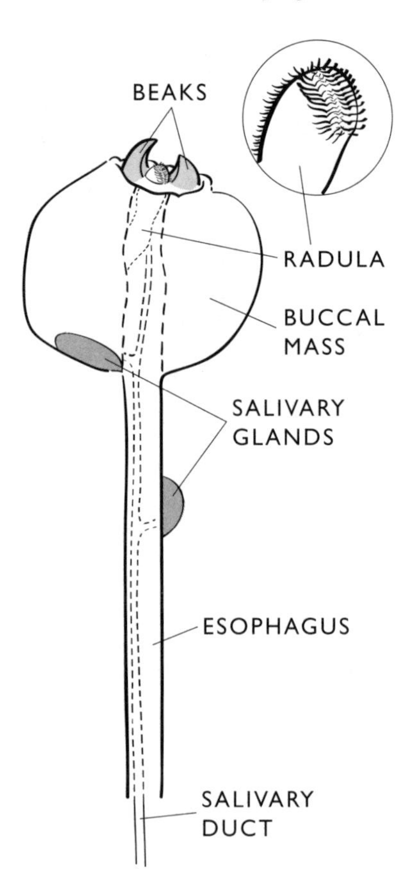

This illustration is an approximation of these tiny parts and their connections; as you'll see in the real thing, the connections between them aren't visible to the naked eye!

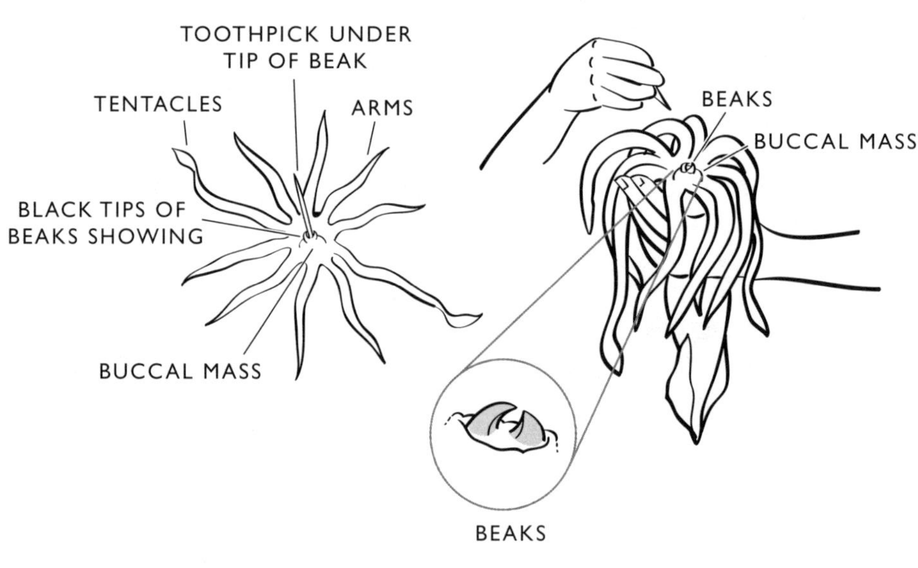

6. Instruct the students to use the toothpicks to remove first one half of the beak and then the other. Alternatively, the whole buccal mass can be pulled out with the beaks inside and then the beaks can be removed (see illustration at left). Attached to the tongue between the two parts of the beak you may see what looks like a toothed ribbon. This is the **radula,** and it shreds the pieces of food torn apart by the beaks. As the buccal mass (which contains the beaks, the tongue, and the radula) is removed, you can see a long tube attached to it. This is the **esophagus,** which passes through the center of the brain and connects at its other end to the stomach. You may also see a small knob connected near the top of the esophagus, and possibly two more knobs a little lower down. These are the **salivary glands,** which secrete mucus and digestive enzymes and—in some other cephalopods, at least—toxins.

Sometimes, when pulling out the entire buccal mass with esophagus attached, you'll also see a slender, string-like structure dangling from the end of the esophagus. This is the salivary duct, which carries digestive enzymes to the stomach through the esophagus (which passes through the brain).

7. Have the students look at the large **eyes.** Explain that they are structurally much like our eye, with a cornea, iris, pupil, lens, retina, and optic nerve. The lens, however, is shaped quite differently from ours, not quite round. Explain how to carefully snip open the eye and use the fingers to search for the lens (it's the only hard part of the eye, and easy to feel). Look at it with the hand lens or under the microscope. Explain that the squid can tell the difference between light and dark but probably can't distinguish colors. It can see a complete image of whatever it's looking at (unusual in invertebrates), and has well-developed depth perception. These adaptations help it capture prey and escape predators. Squids can also perceive polarized light; this may allow them to see "through" the camouflage of some shiny-sided fish, whose silvery scales act like mirrors.

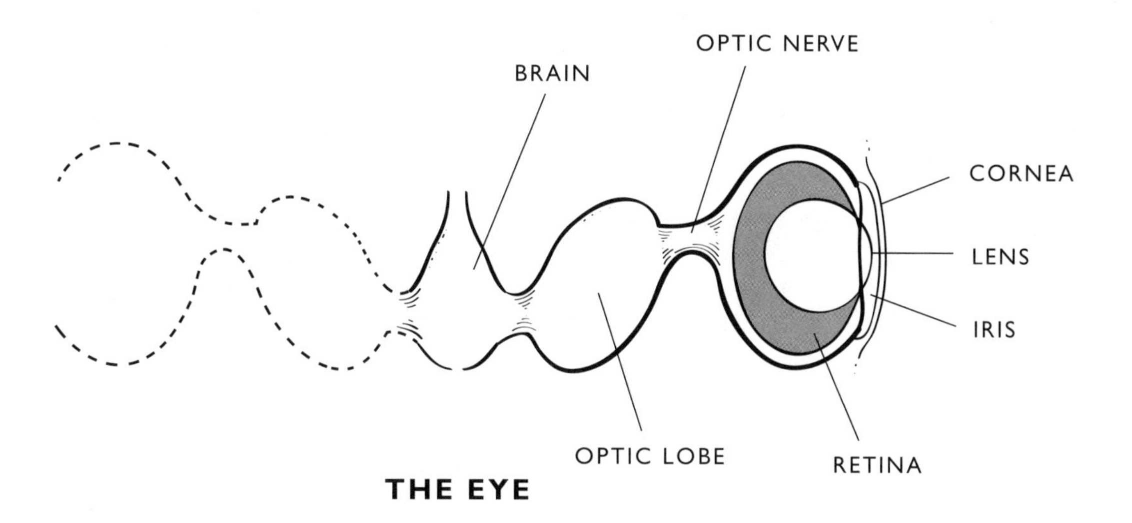

THE EYE

8. At this point in the external examination, put this question to the class again: "Why does the word ***cephalopod,*** or "head-foot," describe the squid so well?" Think back: The eight arms and two tentacles emerge from the head. The beaks are in the head. The eyes are in the head, and the brain is in the head between the eyes. The funnel is attached to the head as well as to the mantle. (Funnel and mantle are described below.) "Head-foot" is definitely the right idea!

External Anatomy: Mantle, Color, Fins, and More

1. Next have the students look closely at the main part of the body. This dotted and colored tube surrounding all of the internal organs is called the **mantle.** Have the students use their hand lens and describe all the colors they see in the small dots. [Brown, beige, black, red, purple, and orange are some of the possibilities.] Be sure to add all these dots of color to your squid chart.

2. Explain that these small dots covering the mantle are pigment cells called **chromatophores** (color carriers), each surrounded by muscles. The muscles are under nervous-system control, which means that the squid can rapidly change colors based on whether it's angry, scared, hungry, courting, or hunting. In this way it's able to distract enemies, camouflage itself to match the background, attract mates, and communicate. When the squid's brain sends a message to change color, the muscles surrounding the chromatophores contract to pull open the sacs. Each chromatophore is stretched out to reveal the pool of pigment within, and the total effect is that color appears to spread out over the surface. The animal darkens in color. When the muscles relax, the chromatophores become smaller and the animal appears lighter. If the squid needs to blanch (become paler than normal) the muscles around the chromatophores relax, which restricts the pigment and makes the animal look pale.

3. Ask the students if they notice a difference in color between the upper (dorsal) and lower (ventral) sides of the squid. Tell the students that squids, like many open-ocean animals including fish, seabirds, and whales, are lighter on their underside or belly and darker on their back. This is called **countershade coloration.** Explain that

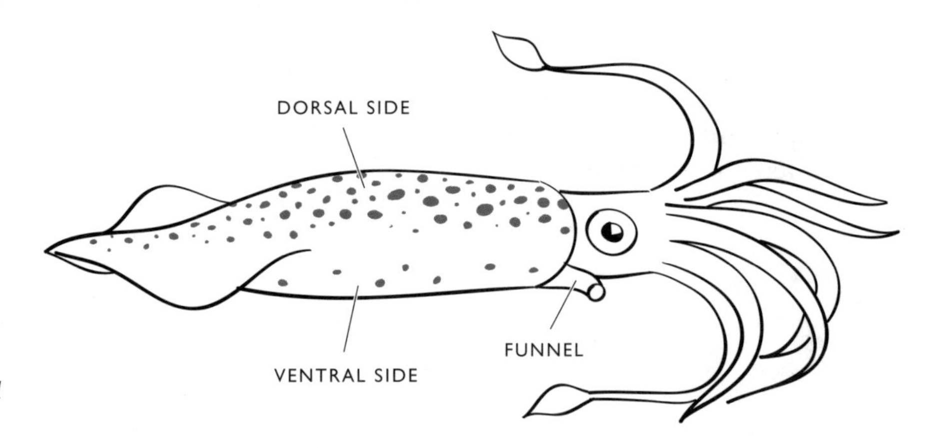

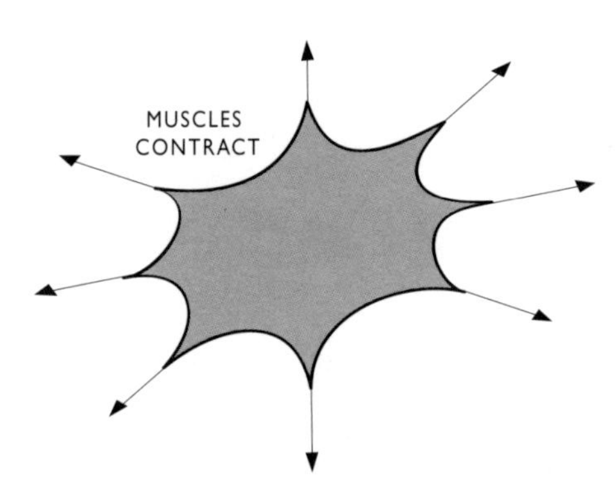

CHROMATOPHORE EXPANDING...

...AND CONTRACTING

MUSCLES RELAX

Different-colored chromatophores are controlled by different nerves. The squid can selectively contract or expand certain chromatophores to achieve a particular color. By expanding all of the red chromatophores, for instance, the squid can flush deep red...something it often does when excited. For animated examples of chromatophores in action, see http://hermes.mbl.edu/publications/Loligo/squid/skin.0.html. (Don't miss the fifth screen, which effectively demonstrates how muscles contract around each chromatophore.)

if a predator hunting for a meal is below the squid and looks up, it sees the lighter underside of the squid, which looks like and blends in with the sunlit portion of the ocean above the prey. If the predator is above the squid and looks down, it sees the darker side, which blends in with the dark color of the deeper ocean below. Counter-shading also helps the squid when *it* hunts; its prey doesn't see it from above or below. Even as a mid-water species without hiding places, the squid is camouflaged from both predators and prey—it "hides in plain sight"!

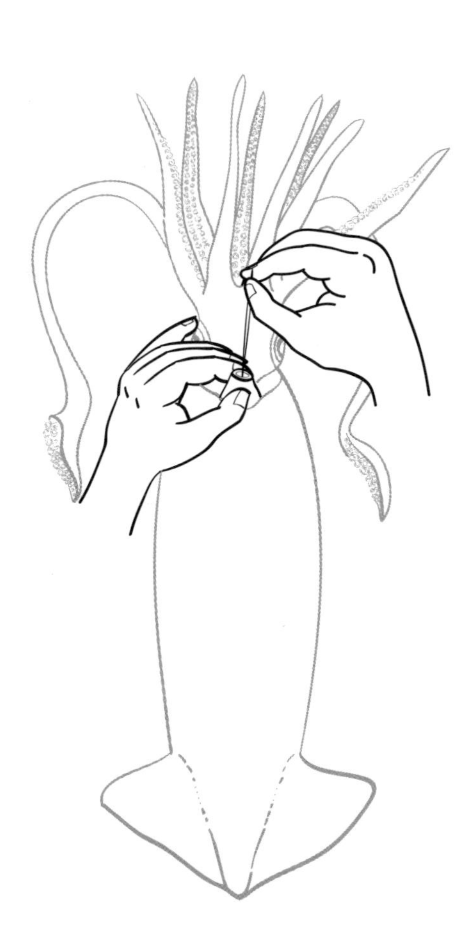

4. Ask the students to recall the video. Did they see the squid using its **fins** to swim? Now have the students locate the two fins on the mantle near the pointed end ("tail end") of their dissection squids. Explain that these fins are used as stabilizers and to propel the squids with dainty motions at relatively slow speeds.

5. Next, describe and show on the chart where to find the **funnel** (siphon). This is a short tube with one opening below the eyes and the other end just under the mantle opening. Have the students stick a toothpick in one end of the funnel and see it come out the other. (Students may miss the opening of the funnel near the eyes. It can be seen most easily if the funnel is carefully pinched at the upper end to open it up.)

6. Describe how the funnel works to propel the squid through the water in the opposite direction from which the funnel is pointing. Thanks to this jet propulsion, squids are the fastest-swimming invertebrates in the ocean.

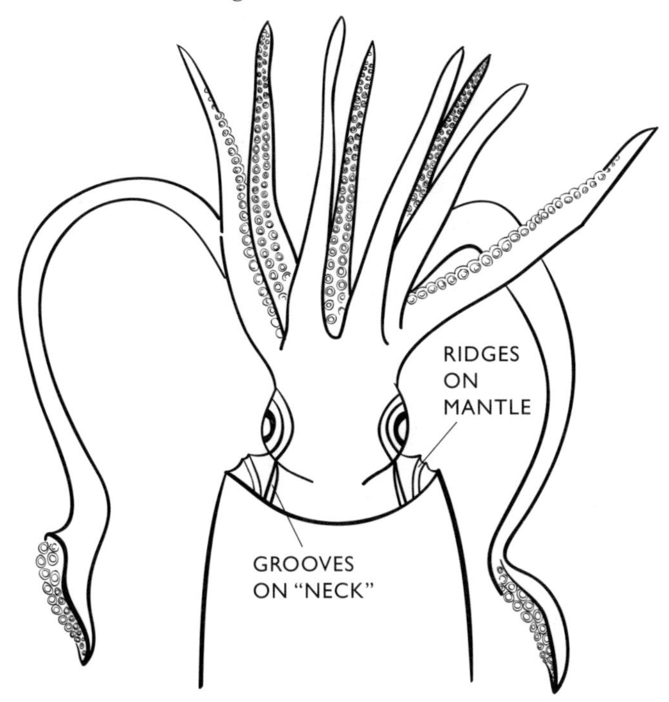

Jet propulsion in squids works like this: The squid takes a lot of water into the mantle cavity through the large opening around the front of the mantle. The muscles close off the mantle opening by "zipping up" two ridges, one on each side of the interior of the mantle, to corresponding grooves on each side of the funnel. Now the only way the water can leave the mantle cavity is through the much smaller diameter of the funnel. The mantle muscles contract and the water jets out of the funnel with enough force to propel the squid through the water at about 7 feet per second, or 20 miles per hour!

7. Hold up your squid and ask students to describe the shape of the body. Explain that the squid is shaped very hydrodynamically: its shape, like a submarine's, creates very little resistance to the water. It can therefore attain rapid speeds to escape predators and capture speedy prey. The students probably noticed that the animal is shaped to cut through the water as it's propelled backwards. Explain that it doesn't always go directly backwards when propelled by the funnel because **funnel retractor muscles** can direct the flexible funnel in any direction. We'll see these muscles when we cut open the mantle.

8. If needed, give the students time to finish their drawings of the external anatomy.

9. Now have students use a ruler to measure the **mantle length** of their squids. They should measure along the dorsal side (the side opposite the funnel), from the pointed tip, or "tail," to the front edge of the mantle. (The measurement stops here, not at the beaks. The head is pretty stretchy, and including it would make the length too variable.) Have them record the size on their paper.

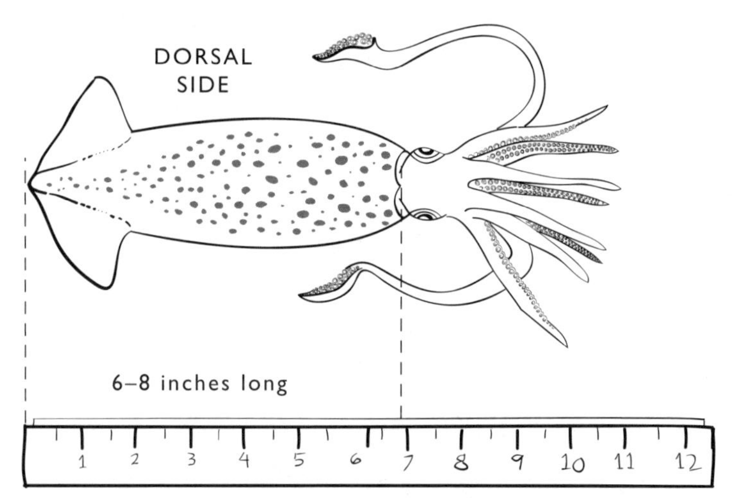

We recommend that you use the same squids for the external and internal observations. Especially if it's your first time doing the dissection, it may take you and your class 1–1 1/2 hours to complete the dissection. If you need to break the activity into two sessions, the squids can easily be stored in labeled plastic bags in the refrigerator. Teachers have also commented that having extra adult help makes the dissection move along more quickly and smoothly.

10. Tell the students that they'll now look at the *internal* anatomy of their squids. Before they use the scissors, though, have students take a peek inside the mantle through the mantle opening and then, as a class, hypothesize about what they'll find inside. List the vocabulary as they make predictions.

Internal Anatomy

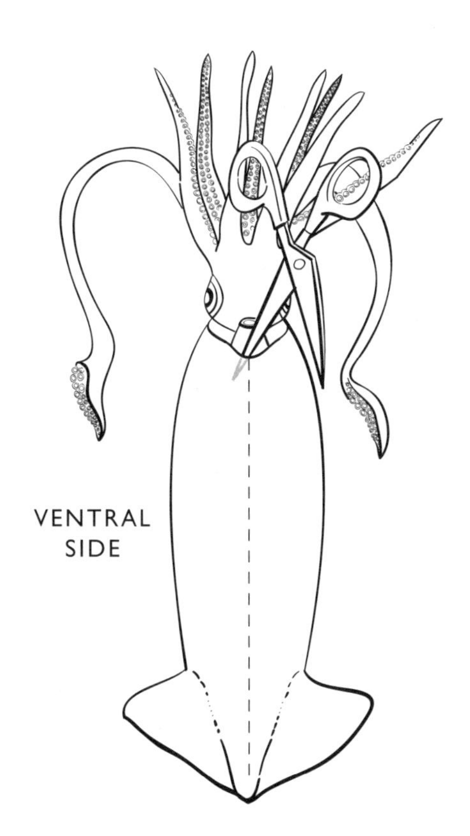

1. Reorient your squid on your plate so that the lighter (ventral) side and the funnel are both facing directly up and both fins are flat against the plate.

2. Be sure the students' squids are oriented correctly on their plates (lighter ventral side up). This ensures that everyone cuts the squid open in the same manner, and that no vital organs are cut through.

3. Distribute a second sheet of blank paper to each pair of students. Have them label it "Internal Squid." Remind the students that as each part of the internal anatomy is presented, you will add the corresponding structure and vocabulary to the class Internal Squid chart, and they should do the same on their own charts. This is a good time for them to trade roles—dissecting and drawing.

4. Model how to cut open the mantle of the squid directly down the midline of the ventral (belly) side. Using the scissors, start cutting at the mantle opening next to the funnel and continue down to the pointed tip of the body. **Alert students to be careful to cut just the mantle and not the underlying organs.** (It's helpful if one student holds up the mantle and the other cuts it. Also, the student doing the cutting can lift the scissors a bit so the lower blade lifts the mantle away from the internal organs.) Once the squid is cut open, have the students spread back the sides of the mantle like an open book and compare their squids with those of other students.

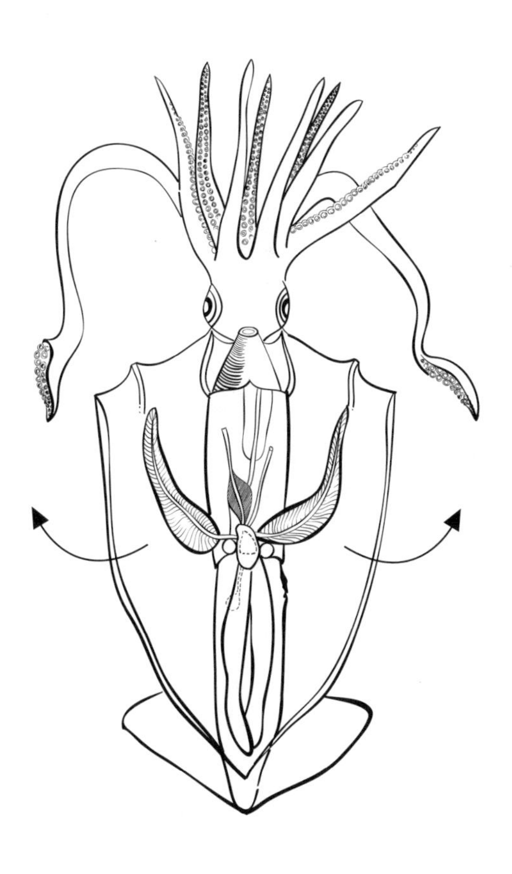

5. Give the students a few minutes to talk about what they see. Then ask them what differences they notice between their squid and others at their table or around the room. How many different "kinds" of squid do they think are represented in the room? [Students may come up with lots of ideas. Some will probably suggest that there are two "main" kinds—males and females.]

Reproductive System: Females and Males

1. The first question asked by students is often, "How do you tell the difference between male and female squids—and which one do I have?" Explain that in sexually mature squids, the **gonads**—the **ovary** containing eggs, in females, and **testes** containing sperm, in males—are located from near the very posterior tip of the body to about the mid-point of the mantle. (Next to the gonads, and about the same size, is a large organ called the **caecum**—this will be discussed under Digestive System, below; it also appears on the Internal Squid Diagram.)

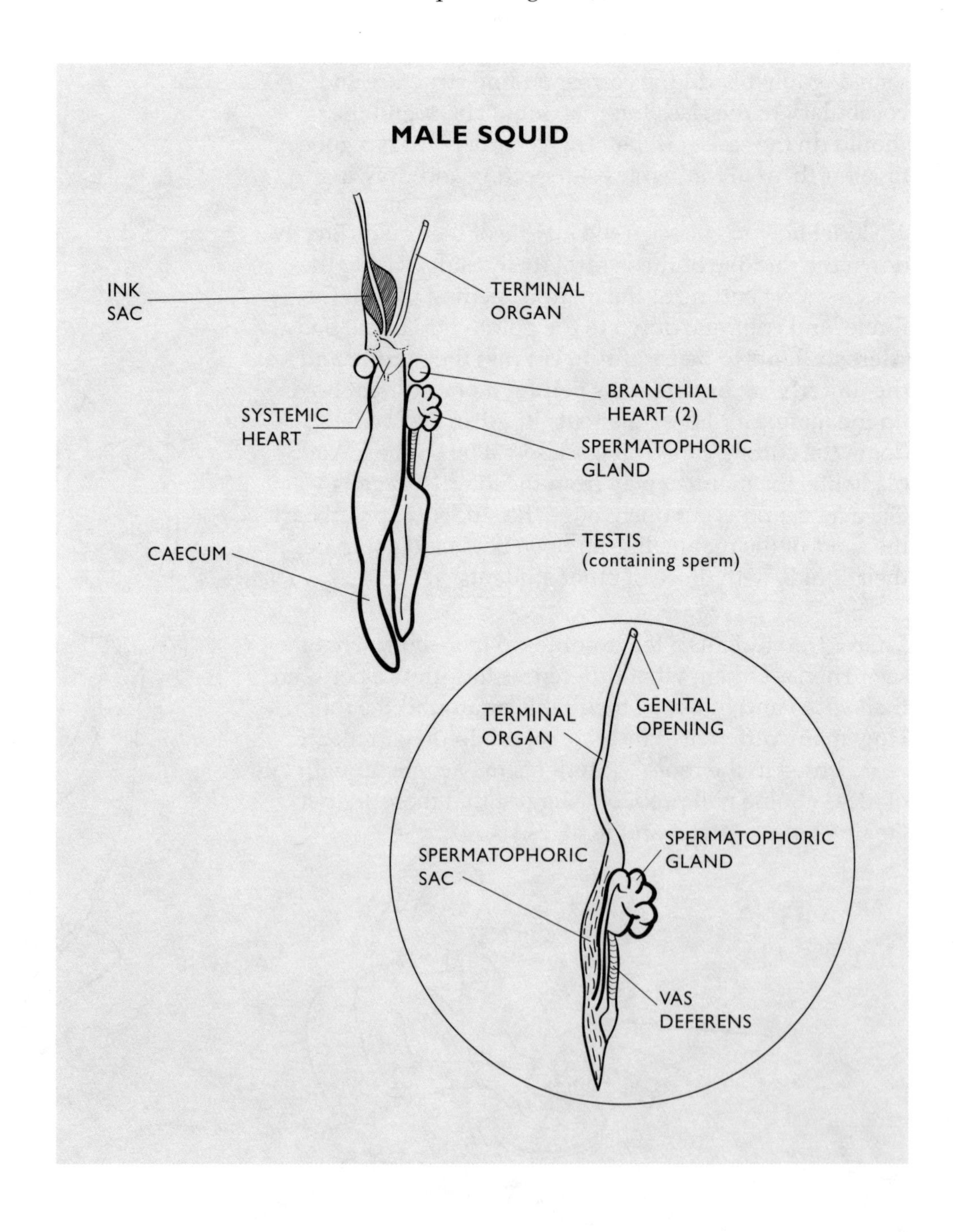

2. Describe the difference between male and female squids as follows:

- Females have **eggs,** which are light yellowish in color and feel like jelly. Depending on the reproductive stage, the eggs may be plentiful and visible as tiny spheres in gelatin (almost like tapioca pudding!), or as a much smaller, more watery gelatinous area. Females also have large white organs in the center of the mantle cavity (body cavity): the nidamental glands (see Females Only, on the following page).
- Males have **sperm,** which is white in color and more cloudy and watery than the eggs.

3. Ask the students to try to determine whether they have a male or female squid, and then circulate around the room and have the students tell you which one they have. Make sure everyone has the opportunity to see both sexes; invite students who find eggs or sperm to share their discoveries with others in the class. (See sidebar.)

Teachers have reported some student frustration in not finding certain organs or structures, especially the reproductive organs. It's not uncommon for students to have difficulty locating certain structures—don't let them be discouraged! The quality, and sometimes even the species, of squid may vary (especially during an El Niño *year, when weather disruption includes a rise in sea-surface temperature and a drastic decline in oceanic productivity). Also, the squids must be sexually mature for all the reproductive organs to be distinguishable. And if the squids are quite large, they may already have spawned—which changes the appearance of the testes and ovaries.*

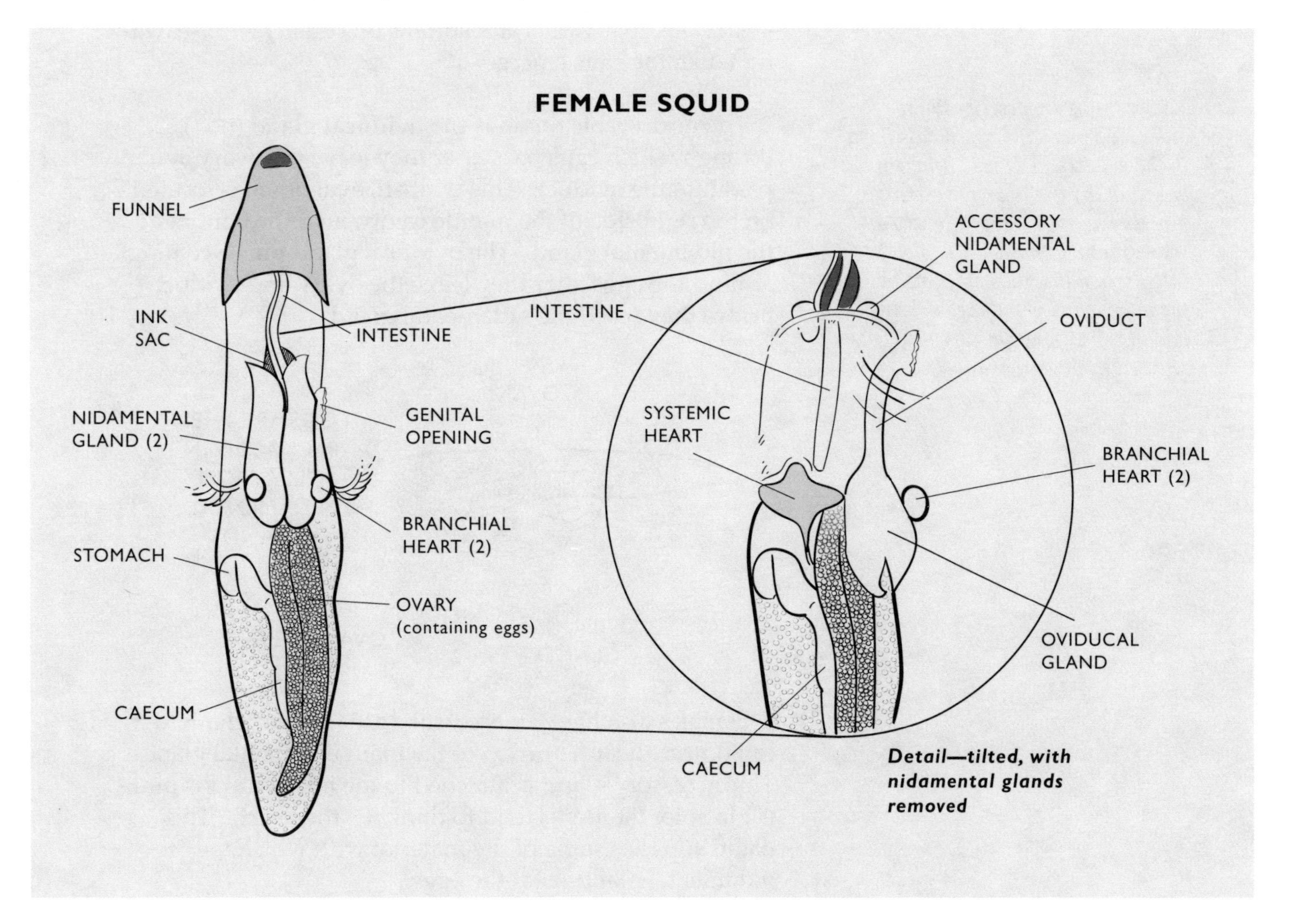

The size and sex ratio of squids found in most samples can be explained as follows: Females come together in huge ***shoals*** *to lay their eggs. Large males swim into the center of the shoal to mate with the females and then jet away. Males that happen to be in the shoal at the moment the nets are set are caught with the females. (At this time, it isn't known why there are almost always just a few small males. In some species these are the "sneaker males," which try to mate with the females while the larger males are busy displaying to one another. Also, squids can be cannibals… smaller males are likelier to be eaten!)*

4. Tally up how many squids of each sex are present in the room, and compare the size ranges between the males and females. Although it's sometimes dependent on the time of year, usually you'll find that the majority of squids will be females of all sizes, and the next most numerous will be large males. Small males are the least common.

5. Depending on how in-depth you've decided to describe squid reproduction, continue with the descriptions that follow ***or*** skip down to the Note on Squid Reproduction on page 65, which describes "how they do it" without using a lot of vocabulary or going into extensive detail.

Females Only

1. First describe the female system. The first reproductive organs the students will encounter are the **nidamental glands** (see shaded sidebar). These are large, oval, white organs located at about the midpoint of the mantle cavity. Explain that the nidamental glands put a protective jelly coating on the eggs just before they're laid. This protective gelatinous coat swells and stiffens on contact with seawater when the eggs emerge.

These organs are described in the order students will come across them in the dissection. It may help you to be aware that, in the reproductive process, the eggs leave the ovary and travel through the oviducal gland into the oviduct, then are coated by the nidamental glands before being laid in egg strands, or capsules, released through the funnel.

2. The next visible organ is the **oviducal gland** (no *t*), through which eggs pass after they leave the ovary and pass into the oviduct. This whitish, oval organ is located on the right side of the mantle cavity, near the bottom of the nidamental gland. The oviducal gland puts a coating around the eggs after they leave the ovary and oviduct, before they reach the nidamental glands.

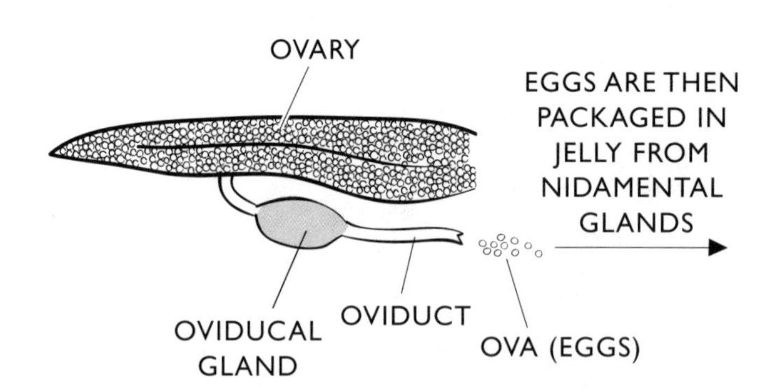

3. Females also have an **accessory nidamental gland** located near the top portion of the main nidamental glands. This accessory gland is attached to the ink sac and is pinkish in color (students tend to think it's the heart). This gland supplies some of the material with which the nidamental glands coat the eggs.

Males and Mating

1. Next describe the male system. The sperm leaves the testes and makes its way to an area called the sperm bulb. From here, the sperm passes through the coiled tube called the **vas deferens** and into the **spermatophoric gland,** which appears as a small sac with many intertwining circles within it. This gland adds coverings to the sperm to make it into a spermatophore, or sperm packet.

2. From the gland the spermatophores exit to the seminal vesicle, or **spermatophoric sac,** where they are stored. From here the spermatophores travel up the **terminal organ** and into the mantle cavity.

*Squids have no real "penis"; their **terminal organ** is not inserted into the female.*

Squid Reproduction Wrap-Up

If you're using the Monterey Bay Aquarium video, turn on the squid footage again (without sound), showing reproduction, egg laying, young hatching, and adults dying. As the class watches the video, explain that reproduction takes place as shoals of squids gather in shallow water for mating.

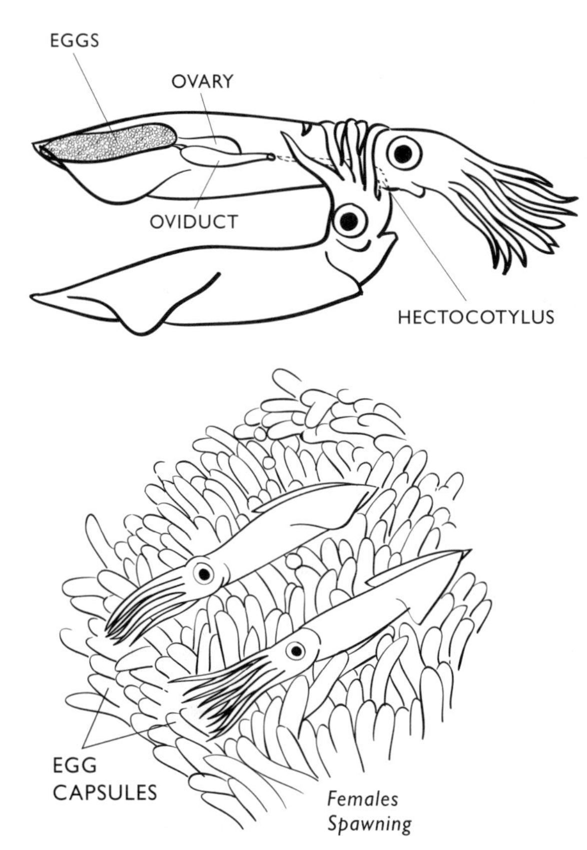

Note on Squid Reproduction: *Males swim alongside the females and grasp them near their heads with their arms and hold them in close (refer to the video footage). The males then reach into their own mantle cavity with a modified arm (called the hectocotylus), scooping up a sperm packet and placing it within the mantle cavity of the female or in a pouch-like structure under the female's mouth. The sperm fertilizes the eggs while they're still soft and penetrable, the nidamental glands add their protective coating, and the female expels the eggs (spawns) through the funnel. The egg capsules—strands of about 180–300 eggs, each strand about the size of your index finger—are glued to plants or rocks on the ocean bottom. Each female lays up to 20 capsules per day for three days and then, like the males, dies. The young squids hatch in a month or so, looking just like miniature adults about 1/10th of an inch long.*

Digestive System

1. Explain that the digestive system in the squid starts with the beaks and radula within the buccal mass, where digestive juices from salivary glands are added, and then continues down through the esophagus to the stomach. Because the esophagus passes through the center of brain, squids must be careful to tear off and swallow *small* bites, to prevent fragments of food or shell from getting stuck in or damaging the brain as they pass through.

At this point it's usually interesting to relate squid anatomy to humans. Students usually know about their own digestive system.

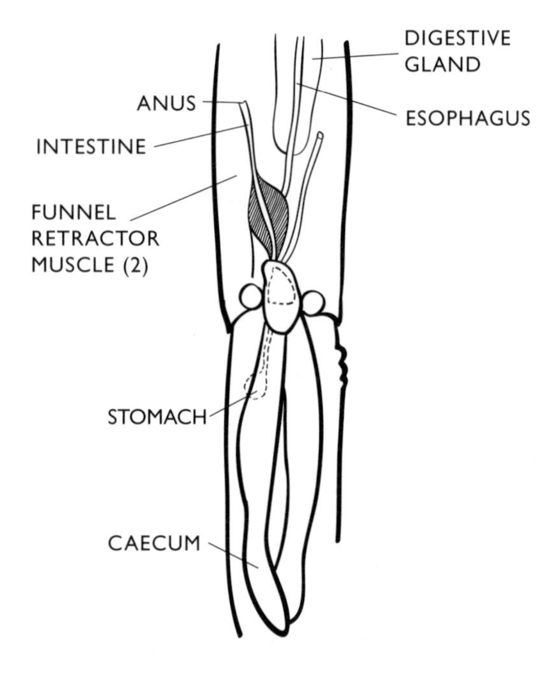

2. Have students locate the **stomach.** It's an oval structure (sometimes difficult to find), about one-half inch long, hooked to the side and near the top portion of the **caecum.** The caecum lies next to the gonads, and is often about the same size and shape. (The stomach and caecum often have little pieces of food in them; looking for these pigmented bits may make it easier to locate those organs.) The stomach is the major site for digestion and the caecum increases the surface area for digestion of food.

3. Continue with the description of the digestive system. Waste products leave the caecum by way of the **intestine,** a long tube emptying through the **anus.** The anus opens into the funnel; as water is released during jet propulsion, the waste products are also jettisoned out of the body.

4. Have the students recall that the funnel can be pointed in different directions. This is the job of the **funnel retractor muscles.** These muscles, along with cephalic (head) retractor muscles, attach the funnel and head to the mantle and feel like strong tendons—a much different consistency from the rest of the squid's soft body. These muscles are attached to the mantle along each side of the diffuse, yellowish **digestive gland,** which provides digestive enzymes to the stomach and caecum.

The Ink Sac

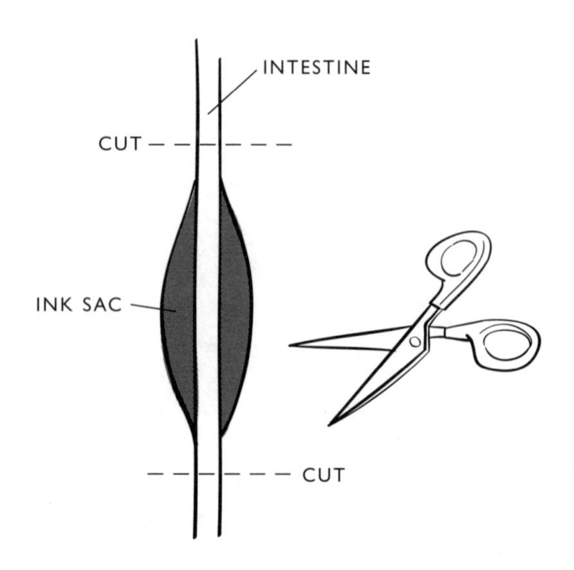

1. Have students visually locate the **ink sac.** It's attached to the intestine and can look very much like a small, silver fish or sometimes like a thin black line, depending on how much ink it has inside.

2. Have the students carefully remove the ink sac by snipping just beyond both ends with their scissors. **Ask them to take care not to puncture or cut the ink sac itself.** The ink is pretty messy, and could make it difficult to distinguish all the parts of the squid. (If the sac *should* get punctured, or come that way from the market, the squid can be carefully rinsed off.)

3. Have the students carefully place the ink sac on their plate next to the squid. Tell them to keep track of it, because the class will return to the ink sac later.

4. Ask the students if they know what purpose the ink sac serves. After listening to their ideas, explain that the function of the ink is to camouflage the squid when a

predator is chasing it. The black ink is jettisoned out the funnel in a squid-shaped blob, which acts as a decoy to confuse the predator while the squid changes to the color of the water and jets away. The ink is also thought to anesthetize (numb) the predator's sense of smell, which makes it more difficult to pursue the squid.

Although each action uses a different contracting (sphincter) muscle, the funnel could probably propel a squid through the ocean, discharge waste products, and aim a cloud of ink at a predator all at the same time. How efficient would that *be?!*

5. Place an ink sac—the one from your squid, or an extra—in a small jar of water and pierce it. (Don't use one of the students' ink sacs, because they'll need their ink to write with later.) Watch how the ink, which contains the pigment melanin, diffuses through the water. (The ink from cuttlefish, a close relative of squids, formed the brownish sepia ink used by artists for at least 2,000 years—long before India ink!)

Many people consider squid ink to be a delicacy. Some gourmet food stores even sell squid-ink pasta, which looks like black noodles.

Respiratory and Circulatory Systems

1. Review with students the fact that water contains oxygen, which marine animals use to breathe. Ask the students what part of the squid anatomy is used to get oxygen from the water. [The **gills.**] Have them locate the gills, which look like two white, feathery structures located along each side in the mantle cavity. Explain that water entering the mantle cavity to be used for jet propulsion also serves to bathe the gills with oxygenated water.

2. Now ask the students what part of the squid anatomy is used to pump blood throughout the body. [The heart—but not just one. Squids actually have three hearts!] Show where the two **branchial hearts** ("gill hearts") are located at the base of each gill. The third heart, called the **systemic heart,** is larger and located between the two branchial hearts. (The kidney may overlay these organs; if you have trouble finding the hearts, use the toothpicks under very good light to tease away some of the kidney tissue.)

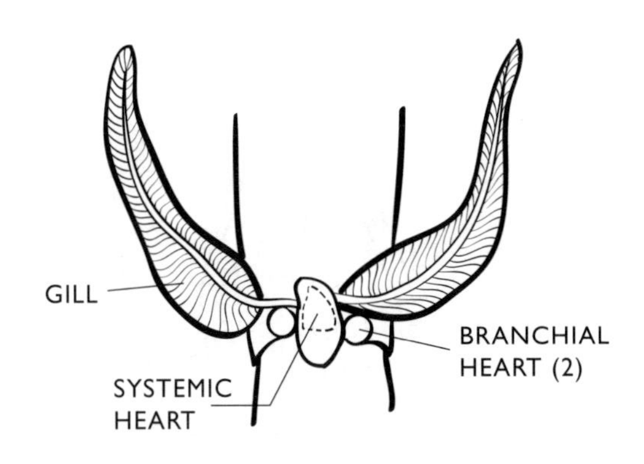

The two branchial hearts pump blood from the body to the gills to be oxygenated. These hearts are analogous to auricles in humans. Each branchial heart is quite small and slightly yellowish in color. (These hearts are more difficult to see in the female because the nidamental gland overlays them. Remove the gland, if you like, for better viewing.) The systemic heart pumps oxygenated blood from the gills to the rest of the body and is analogous to the human ventricle. The blood contains the oxygen-carrying pigment ***hemocyanin,*** *which is blue when oxygenated and colorless when without oxygen.*

Support System

1. Explain that the squid is supported as it speeds through the water by a stiff, slender structure called a **pen,** or gladius, which runs the length of the dorsal midline of the mantle. The pen is made of chitin (like the beaks), and looks much like a transparent feather or thin piece of clear, flexible plastic. This structure is a remnant of the shell that protected the ancestors of modern-day squids, and that still supports and protects other mollusks such as the chambered nautilus (a cephalopod), snails, and clams.

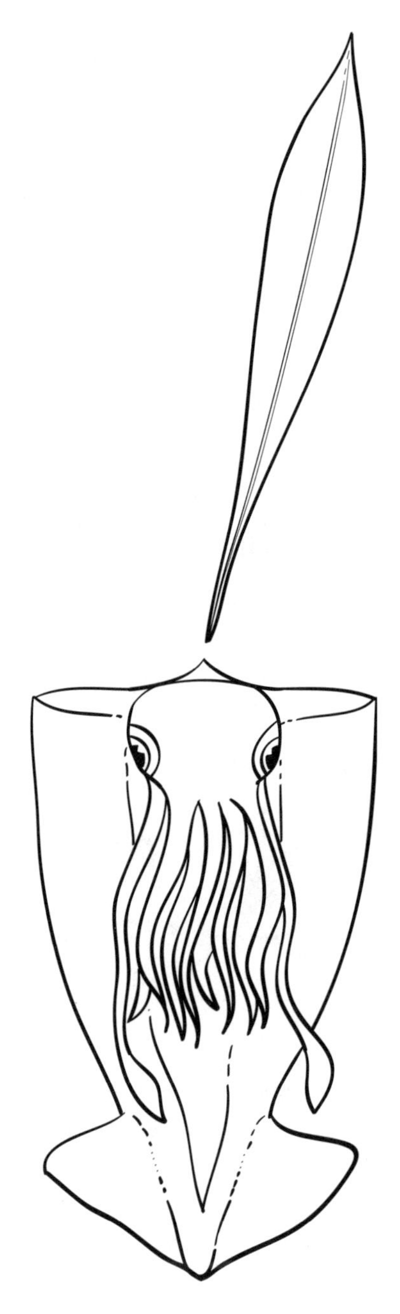

Squids are unique among animals studied in human brain research because their nerve fibers, or axons, are the largest in the animal kingdom: about the width of a pin—100 times larger than in humans! This makes squids excellent candidates for nerve-impulse research, and much current knowledge about nerves has come from studying the squid's giant axon. Studies contributed to at least four Nobel Prize awards between 1963 and 1970 alone! (See "Behind the Scenes" on page 149 to learn more about the squid's giant axon.)

2. Have students locate the pen under all the organs we've been investigating. To find it, lift up the head and lay it down over the top of the organs of the body. Underneath, where the head was lying on the plate, you'll now notice a pointed area touching the plate right along the midline of the body. This is the tip of the pen, covered with thin skin. Scrape away this skin, then grasp the tip and pull, sliding the pen out from the mantle.

3. Have students return to the **ink sac** on their paper plate. They can now cut open the ink sac, tease it apart with their toothpick, and dip their "pen" in the ink to write their names or initials on their plates or drawings.

The Nervous System

1. Let students know that the **brain** of the squid is very highly developed; the squid and its relative the octopus are considered the most intelligent of all the invertebrates. The large brain surrounds the esophagus and lies right between the eyes. It consists of many bundles of nerves fused together and surrounded by a "skull" made of cartilage.

2. Explain that much of the squid's incredible speed is made possible by an unusual set of **nerves.** These are so highly developed with giant nerve fibers, or **axons,** running down both sides of the mantle that when a squid is threatened—or spots a meal—all the nerves react simultaneously to contract the mantle. This forcefully expels the water trapped inside the body and jets the squid away from danger or toward its prey.

3. Have the students cut off the head of the squid and then place the tip of their scissors into the cut portion. Have them carefully and shallowly snip along the dorsal surface between the eyes, explaining that they are cutting through the protective cartilage of the brain case. The brain of the squid will now be obvious.

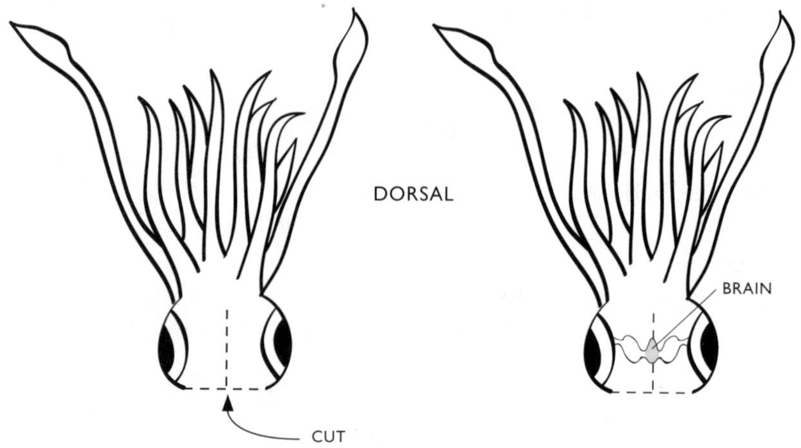

Wrap-Up

1. Collect the students' illustrated internal and external anatomy charts, which will become part of the student portfolios used for assessment.

2. During another period, you may want to revisit the charts to have the students review and write about one of the squid's ***adaptations***—physical structures and how they function to help the squid survive in the open ocean.

The term "adaptation" refers to physical and behavioral characteristics of an organism that help it survive in its environment. Focus on this concept when reviewing the functions of the many squid structures and organs the students observed. The Session 3 activities Squid Pictionary and Squid Jeopardy provide another opportunity to review this creature's fascinating adaptations.

3. If you won't be cooking up the squids, they should be properly discarded. Undissected squids can be returned to the refrigerator and used for another lesson if it occurs within one or two days.

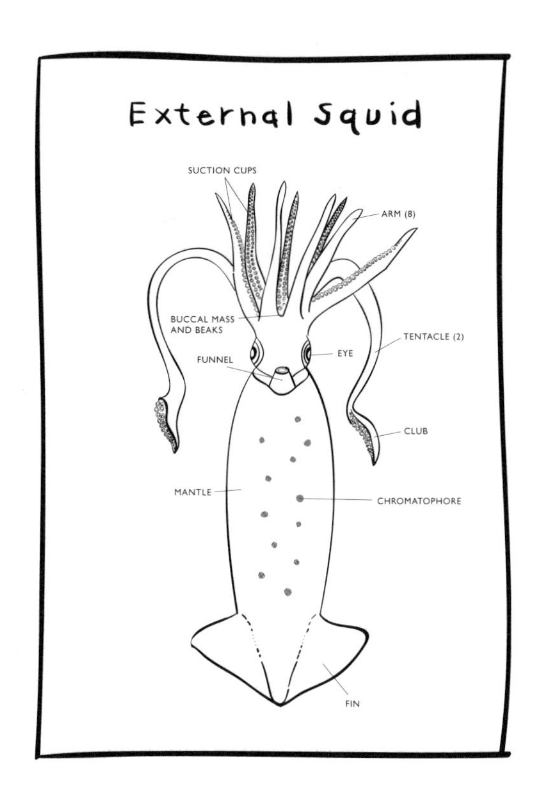

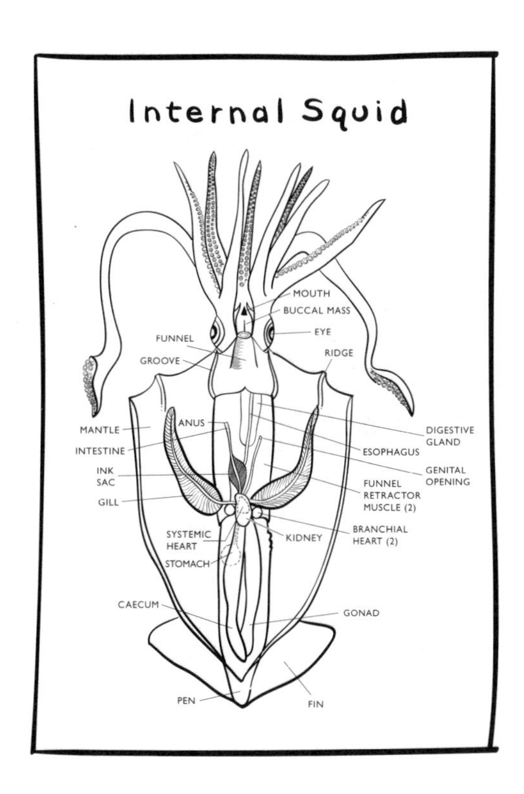

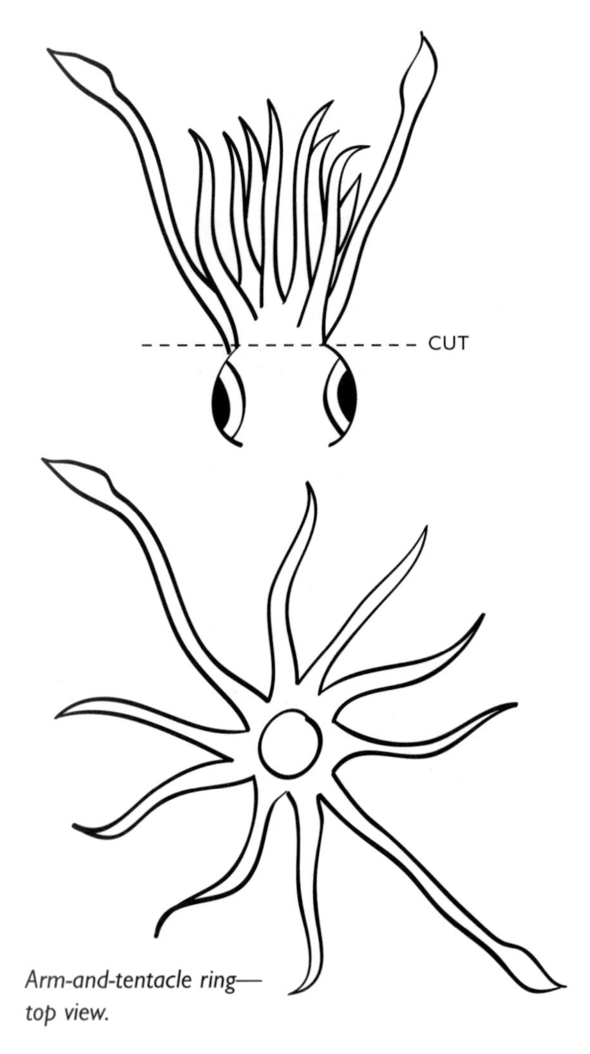

Arm-and-tentacle ring—top view.

Teachers and students rave about doing the Calamari Festival! Fried squid is delicious, and this is a great opportunity to have your students try something new. Clean hands and tools will take care of health concerns; make sure to cook the squids immediately after dissecting, or cook "extra" squids while the dissection is going on. If you're taking the squid home to cook for the next day, keep it refrigerated until preparation.

Calamari Festival

Note: Simple directions for fried squid in tempura batter appear on the next page.

1. Students can now prepare the dissected squid so it can be cooked for the Calamari Festival! (To prepare UNdissected squid, see cooking instructions on the next page.) Describe how to clean the squids as follows: Peel all the colored skin off the mantle, leaving the cleaned white flesh. The arms and tentacles are also eaten; cut them off between the mouth area and the eyes, keeping them joined in one large ring. Decide if you want to let the students keep the pen and the lenses of the eyes, but **do not let them keep any other internal organs.**

2. Have students place the cleaned mantle and the arm-and-tentacle ring on the clean (bottom) plate and have someone collect all the cleaned squids. Have another student discard all the internal organs and skins on the dirty plates into a plastic bag. Collect the scissors to be washed later.

3. Have students clean off their tables, wash their hands, and prepare to eat the calamari (if someone has been cooking it during the dissection).

4. Hold up the first two Key Concepts for this activity and have one or more students read them aloud. Post them near the squid charts for students to refer to later.

- **Pelagic creatures are organisms living in the open ocean.**

- **Looking closely at an animal like the squid can tell us a lot about the adaptations needed to survive and thrive as a pelagic creature.**

If you'd like to try the squids deep fried in tempura batter, follow the directions below.

1. Heat the oil in the "Fry Daddy" or other deep-fat fryer.

If you're using the dissected squids, wash off the mantles and arms with plenty of cold water and pat dry. Slice the mantles and arm-and-tentacle rings crosswise into strips about one-half inch wide, and place into a bowl with the prepared tempura batter (follow the directions on the box).

If you're using squid that was not dissected, follow these steps. Thaw the squid, if it's frozen. Grasp the head and arms firmly and steadily pull on them. Many of the internal organs will be pulled out of the body cavity as the head and arms are pulled off. Place the dull side of a knife at the pointed end of the body and slide it firmly up toward the opening (where the head was). This will squish out any remaining pieces of internal organs. Don't forget to use the arms and tentacles; cut them off the head between the mouth area and the eyes, keeping them all joined in one large ring (crown). Peel the skin off the mantle. Wash the mantle and the arm-and-tentacle crown inside and out with cold water and pat dry. Slice the mantle and arms into half-inch-wide rings and place in a bowl with the prepared tempura batter (follow the directions on the box).

2. Fry the squid for about 30 seconds, until it's just browned (it gets rubbery if cooked too long). Drain on a paper towel and serve with lemon, if you wish.

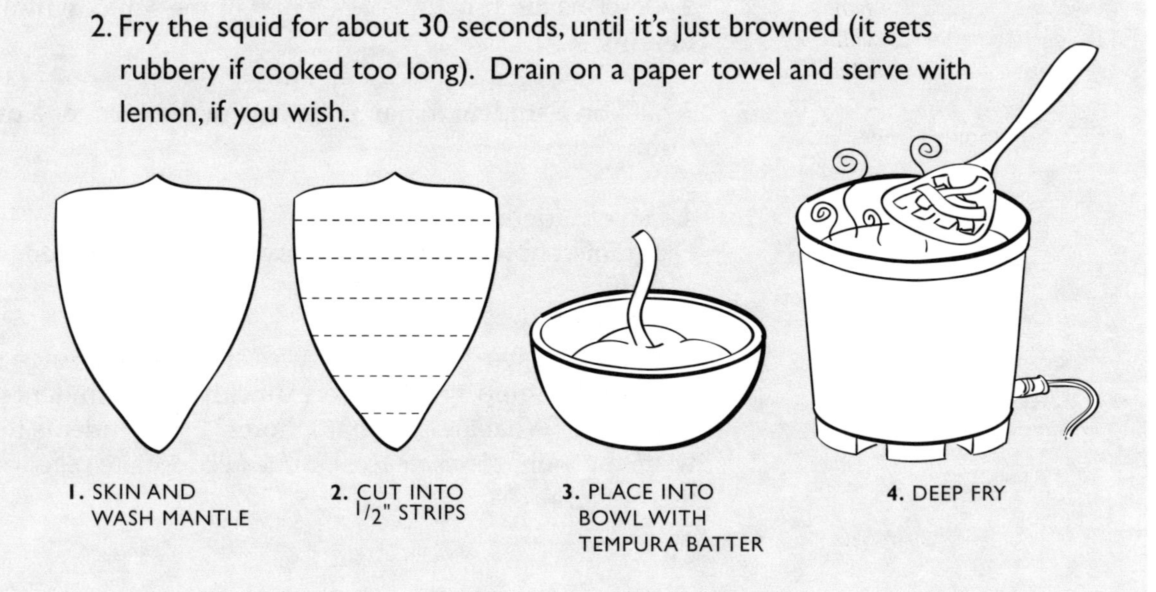

1. SKIN AND WASH MANTLE
2. CUT INTO 1/2" STRIPS
3. PLACE INTO BOWL WITH TEMPURA BATTER
4. DEEP FRY

Session 3: Debriefing Activities

Choose one or all of the following debriefing activities to reinforce concepts, check for understanding, and assess individual knowledge. Keep all the charts and information displayed around the room so students can use them as references.

Squid Pictionary

Students take turns drawing parts of the squid to complete squid chart #3, What We Learned.

1. Write the names of each discussed part of the squid on separate slips of paper and put all the slips in a box.

2. Have a student pick one of the slips and—**without saying the name of the part**—draw it on the What We Learned chart.

3. Allow the rest of the class a fixed amount of time to guess the name of the part and describe its function.

4. Repeat with another student.

Squid Jeopardy

1. Distribute one unlabeled Internal Squid Diagram and one unlabeled External Squid Diagram student sheet to each student.

2. Describe the **function** of one part of the squid, **without naming it.**

3. Call on a student to name the part **in the form of a question.**

4. Have students write the part on their External Squid Diagrams, showing that they know where it's located.

For example:
The teacher says, "These structures are used to change the color of the squid's mantle very quickly." The student responds, "What are chromatophores?" All students then write the word *chromatophore* in the appropriate place on their diagrams.

Video Highlights and Reruns

1. Play the squid video again, without sound.

2. Have students take turns narrating each scene, pointing out parts of the body and describing squids' adaptations to their habitat.

Alternate version:
1. Have students list all the adaptations of the squid as illustrated in the video.

2. Form students into small groups.

3. Assign, or allow students to choose, one listed adaptation per group.

4. Cue up the video to spotlight each adaptation, and have each group make a presentation to the class.

"I Used To Think but Now I Know" Books

1. Have students complete the last column ("What we know to be true") of their individual About Squid Charts.

2. Lead a class brainstorm and write down the students' responses as they share them with the class.

3. Ask students to silently choose their favorite fact about squids and match it with something from the first column of the chart, "What we think we know."

4. Have them write these facts in the form of an "I used to think_______but now I know______" sentence. Have them add an illustration. The individual student pages can be put together to form a book, or used separately in student portfolios.

Session 4: Squid Fishery Symposium

What's My Concern?

1. Have students reflect briefly on what they've learned about squids. Do they appreciate the squid more than they did before? Why? Why do people eat squids? Who are the people involved with catching and eating squids?

2. Tell students that the people who fish for and eat squids are deeply concerned; the demand for squids is going up while the squid population is going down. What should be done?

3. Explain to students that they'll get to represent the viewpoints of different squid fishery "interest groups" at a Squid Fishery Symposium, and that the main purpose of the conference will be to discuss the problems with the fishery and consider possible solutions.

4. Divide the class into six groups, each representing one of the six different squid fishery interest groups on the Squid Interest-Group Chart. Tape up the chart where everyone can see it, **keeping the questions covered for the moment,** and show the class the names of all the interest groups. Tell the student groups they'll be given a description of the point of view their "interest group" represents. They'll read the description and then answer questions from that viewpoint.

5. Assign, or have each group choose, someone to do each of the following jobs:

- **Reader,** who'll read to the group the "interest group viewpoint" given to them by the teacher.
- **Recorder,** who'll take notes about the group's viewpoint as they discuss it.
- **Presenter,** who'll present the interest group's viewpoint at the culminating Squid Fishery Symposium.

6. Pass out a Squid Interest-Group Profile and one sheet of lined paper to each group. The Reader should read the description out loud to the group, and they should then discuss their point of view as the Recorder takes notes.

Share with the Class

1. After each group has discussed its viewpoint and made notes, have the Presenters from each group tell the rest of the class about their point of view. Rather than just read the card, they should use their own words to paraphrase their viewpoint, or act it out in a mini-drama with help from their group members.

2. Following each presentation, give the rest of the class a few minutes to ask the presenting group any questions they may have.

3. Finally, **uncover the questions on the Squid Interest-Group Chart at the front of the room** and lead a class discussion to fill in the blanks. Any of the students in the interest groups may offer suggestions, and the questions don't have to be answered in any particular order. As answers are given, record them in the appropriate place on the class chart.

SQUID INTEREST-GROUP CHART

	Squid Fisher	Consumer	Restaurant Owner	Sport Fisher	Biologist	Environmentalist
#1 How would we know if squids were over-fished?						
#2 How can we prevent squids from being over-fished?						

4. Hold up the remaining Key Concept for this activity and have one or more students read it aloud. Post it for students to refer to later.

- **Many people depend on squids for food or for their livelihood. More discussion among these people will help create solutions to the problem of diminishing squid populations.**

5. Tell the students that in the next activity they'll learn how squids are caught by different countries, and where the major fisheries (commercial fishing operations) are located. We'll look closely at a major worldwide environmental issue: overfishing (overharvesting).

Going Further

1. More About Mollusks

With your class, study the phylogeny (evolutionary history) and morphology (form and structure) of the Mollusca phylum. Give a brief description of mollusks and the relationship of other mollusks to the squid. You may wish to use the poster "Shellfish/Edible Mollusks" for reference (see "Resources" on page 163).

As with all research, care should be taken to find legitimate information about squids when using the Internet. Some sites look authentic enough, but in fact contain exaggeration and outright fiction—particularly about the vampire squid (wouldn't you know it?).

2. Library or Internet Research

Have students do library or Internet research to learn about one or more of the 400 different species of squid, including the marvelous vampire squid *(Vampyroteuthis infernalis)* and "jumbo," or Humboldt, squid *(Dosidicus giga)*. How is information collected on these possibly rare or otherwise difficult-to-study species?

3. Giant Squid Stories

Relate some stories to the class about historical sightings of "kraken" (squid sea monsters). (See "Resources.") Then have students write their own stories or myths about a giant squid. Stories and myths are usually more powerful if they have some basis in fact; remind the students to use what they already know to be true about squids.

4. Peer Teaching or Big Buddies

Have your students teach the Squid Dissection activity to another grade level or another class. Remind them to think about the teaching strategies they'd use, and what parts of the activity they'd want to share with the other class.

1. Family Recipes

Ask your family if they have any favorite squid recipes that you could cook for dinner. Write down the recipes from your family, or one you tried from a recipe book, and add a few illustrations and facts about squids. Bring your recipe and illustration to class to share with your classmates. All the recipes can then be compiled into a squid cookbook with recipes from all over the world. If your family doesn't normally eat squid, suggest that you go with them to the store, purchase the squid, clean it, and cook it using the recipe you tried at school. You might start a new family tradition with a new favorite family recipe!

2. Go to a Seafood Restaurant

Ask your family to go out to a seafood restaurant with you, and order squid for your meal or as an appetizer to share with them. How was it cooked? Was it as good as when you made it in class? Was it as good as your family recipe (if your family cooks squid)? Did your family try it? Did they like it? Ask to speak to the owner or chef and tell her you're studying squids and the squid fishery in school. Ask her if it's more difficult or expensive to get squid now than it was in the past, and what she thinks is causing these changes. How do you feel about the answers you've heard, and what could you and your family do about it?

3. Go to the Fish Market

Visit your local fish market and watch people buying seafood. What seems to be the most popular seafood item to buy? (Ask the sales clerks which seafood customers buy most.) Do they sell much squid? If you see someone buying squid, ask if he would mind answering a few questions for a school project. Some questions to ask include: How do you plan to cook the squid? How often do you serve it? Have you noticed if the species of available squid has changed over the years? How has the price varied? Would you change your buying habits if you learned that squids were being overfished?" Compile your results and share what you learned with your classmates. (While you're at the fish market, you might buy a fresh squid, take it home, and dissect it for your family!)

4. Visit a Museum or Aquarium

Encourage your family to take a trip to a museum or an aquarium that displays relatives of the squid and other pelagic (open-ocean) organisms. The Smithsonian Natural History Museum in Washington, D.C. has a wonderful permanent squid exhibit; the Steinhart Aquarium, at the California Academy of Sciences in San Francisco, and the Monterey Bay Aquarium a few hours south, both display mollusks and other pelagic organisms. Get on the Internet to find other (possibly local) museums or aquariums with open-ocean exhibits. At the aquariums, look for all the organisms you can find that live their lives in the open ocean. Teach your family what characteristics to look for in pelagic animals. What adaptations do they all share?

5. Search for Giant Squid—at the Movies!

If you happen to have cable television, watch for movies on the Discovery and National Geographic channels about searching for giant squids. You can also find a (less-than-realistic!) giant squid in the 1954 Disney movie *20,000 Leagues Under the Sea*. Watch this movie and see if you can pick out parts of the squid you think are realistic and parts that are far-fetched. Several other movies (not recommended for children), including *Sphere* and *The Beast*, portray giant squids as scary, almost alien predators. Why do you think screenwriters, authors, and directors like to use squids as the villain or monster in their movies and books? What is it about squids that makes them scary? Ask your friends and family what they think is scary about giant squids. Which of their ideas about squids are correct and which are just myths and legends? Discuss their ideas and share with them what you know to be true about squids. Check out the resource book *The Search for the Giant Squid* by Richard Ellis. It's a wonderful compilation of facts, stories (real and legendary), books to read, illustrations, and museum displays about giant squids.

6. Make a Squid Costume

Make a squid costume and wear it for special events during Ocean Month, if your school holds one. You could even wear it for Halloween! There's a wonderful squid-costume pattern in the activity book *Sea Searcher's Handbook* (Monterey Bay Aquarium Foundation, 1996).

External Squid

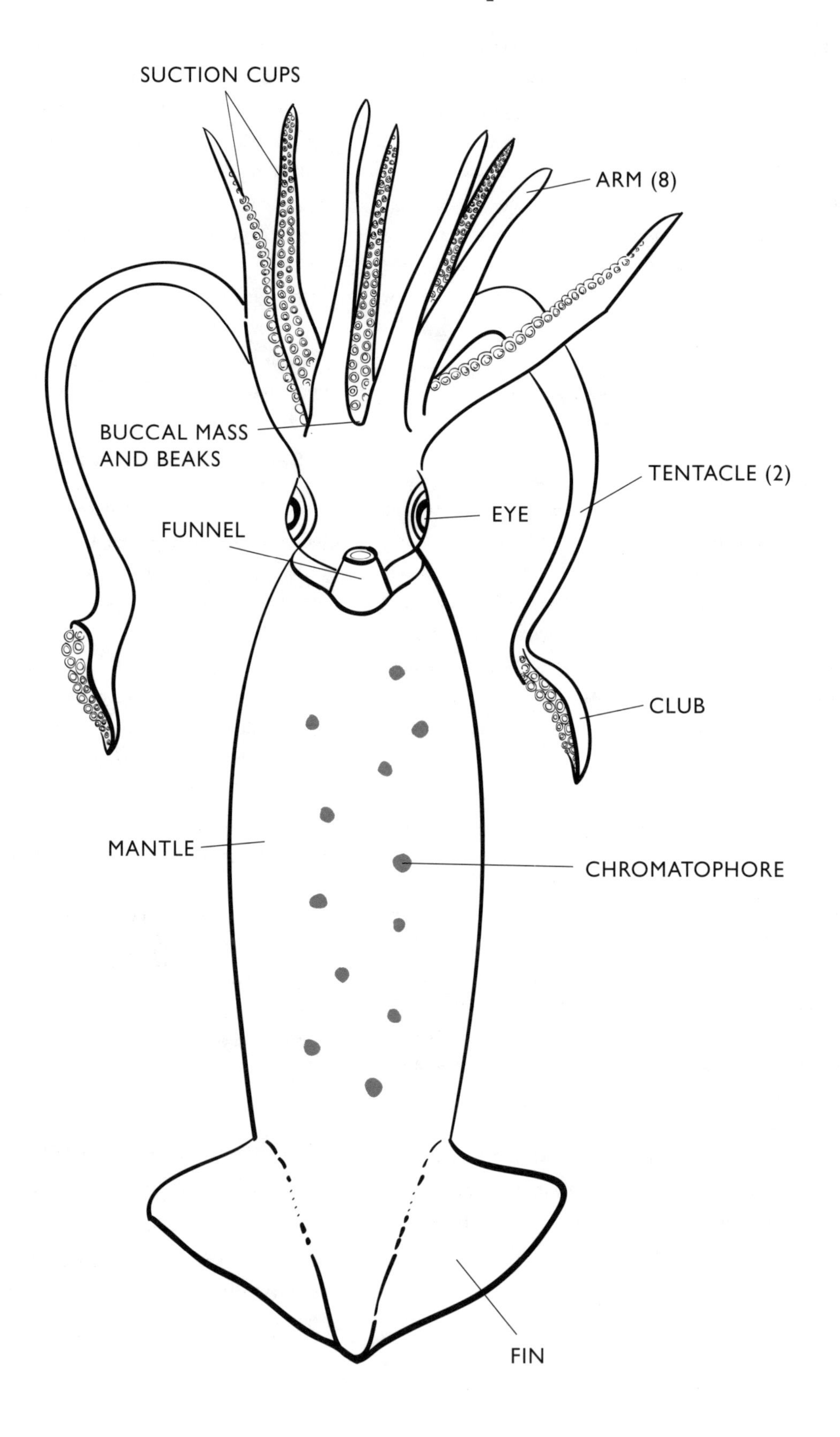

Internal Squid

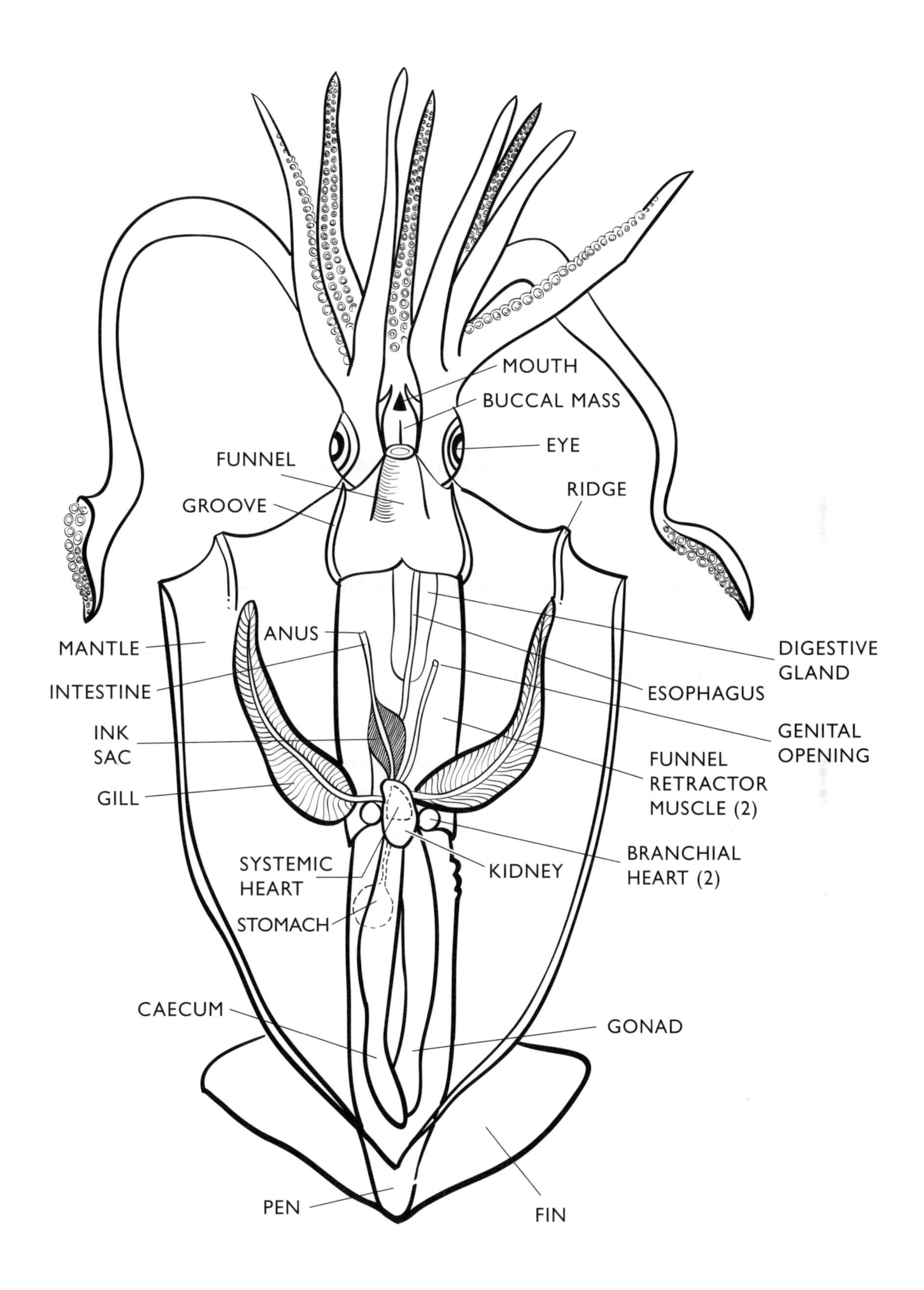

❶ *External Anatomy*

- Arms, Tentacles, and Suckers
  How squids capture prey

- Head
  beak(s)
  buccal mass or cavity (mouth)
  radula
  esophagus
  salivary glands
  How squids eat
  eyes
  lens
  How squids see light, colors, images

- Mantle
  chromatophores
  How squids change colors; camouflage; countershade coloration

- Fins
  Fins as stabilizers

- Funnel
  How funnels work; jet propulsion; retractor muscles

- Length

2 *Internal Anatomy*

- Under the mantle

  What distinguishes males and females

- Reproductive System
  gonads
  ovary
  testis (testes, pl.)

Female system	Male system
nidamental glands	vas deferens
oviducal gland	spermatophoric gland
accessory nidamental gland	spermatophoric sac
	terminal organ

  Mating (video)

- Digestive System
  stomach
  caecum
  intestine
  anus
  funnel retractor muscle
  digestive gland

- Ink sac

  Escape response

- Respiratory and Circulatory Systems
  gills (2)
  branchial hearts (2)
  systemic heart (1)

- Support System
  pen

  (Open ink sac; use "pen" to write their names with ink)

- Nervous System
  brain
  nerves
  nerve fibers/axons

Squid Statements Anticipatory Chart

Statements About Squids	*YES*	*NO*	*Don't Know*
Squids don't change their color			
Squids are usually solitary			
Male and female squids look the same on the outside			
Squids have 10 arms that all look alike			
Squids use fins for locomotion			
Squids chase and catch fish			
Squids lay single, individual eggs			
Young squids look just like their parents, except smaller			
Squids jet through the water with their arms trailing behind			

External Squid

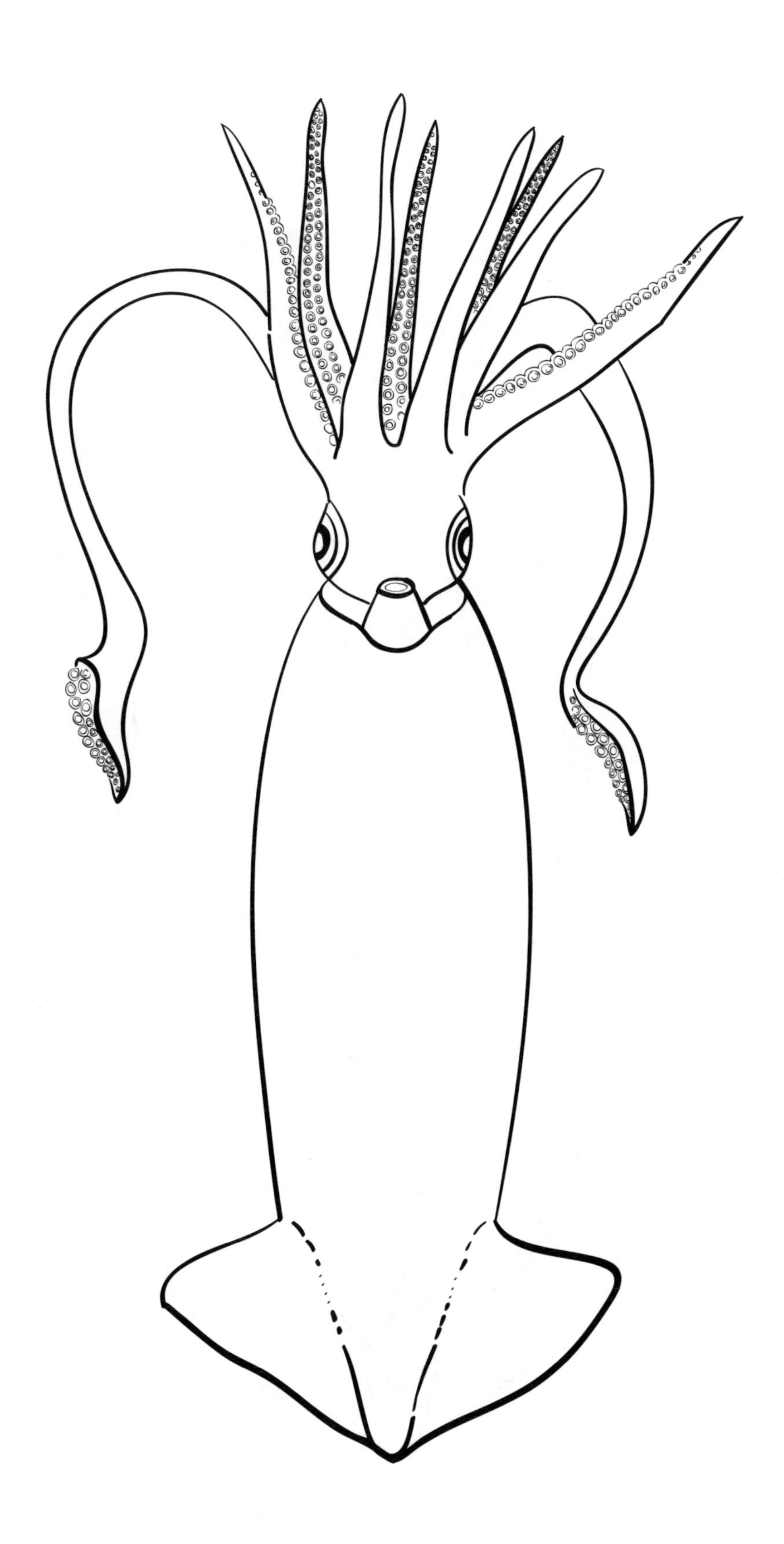

Internal Squid

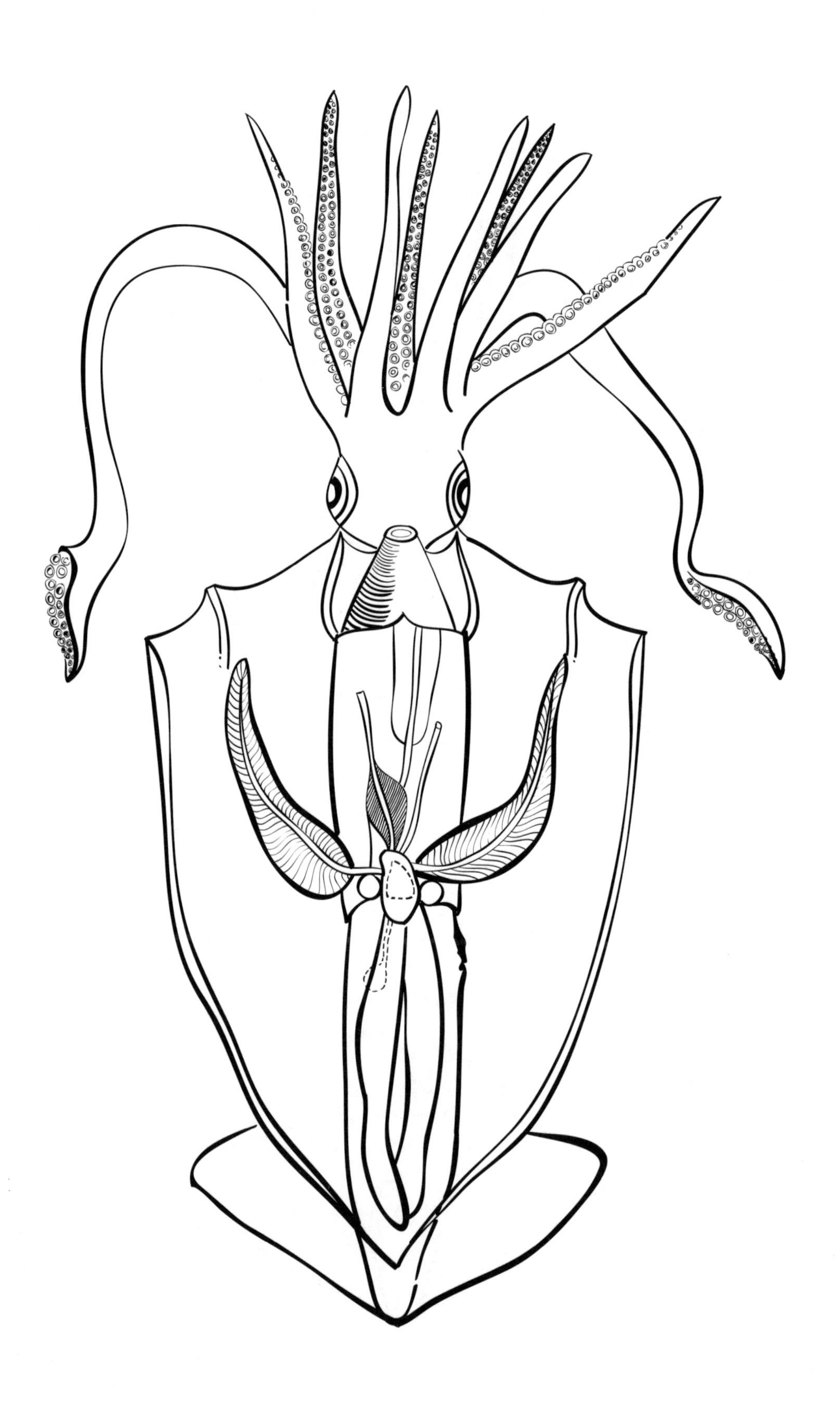

Squid Interest-Group Profiles

Squid Interest Group
Squid Fisher

I've spent my life fishing for squid along the California coast. My boat used to be one of just a few squid boats in Southern California—now there are hundreds. I'm proud of what I do. I work hard, and my catch helps to feed people. I love being on the ocean. I own my boat and support my family with the money I make fishing. My dad and grandfather were fishers too, but they didn't fish for squid; they fished for salmon and halibut. Those fisheries are now closed to all but a few boats each year because most of the fish are gone. A lot's changed since my dad and grandfather were fishing—including the fact that now there are women like me out there hauling nets. I have two kids, and I want them to go to college. I don't know what I'd do if the squid fishery collapsed, because there aren't any open fisheries left in California to go into. Maybe they should limit the number of boats that can fish for squid—as long as they don't limit me! I don't know what else I'd do if I couldn't fish. Don't I have a right to support my family and make an honest living?

Squid Interest Group
Sport Fisher

I have that bumper sticker on my truck, "A bad day fishing is better than a good day working." I've always loved to fish, especially on the ocean. There's nothing like spending a day on a boat. I don't catch enough of anything to damage any whole populations of fish. Sport fishing doesn't need to be regulated. It's those big commercial boats that do the damage, the ones that catch tons of fish. I'm worried about how big the squid fishery is getting. It seems like every time I go out, there are more and more squid boats out there. Some of the fish I like to catch feed on squids—if people overfish the squids, those fish might go away.

Squid Interest Group
Consumer

I really like calamari. I order it whenever it's on the menu at a restaurant, and sometimes I even cook it at home. It's not expensive, it's healthy, my kids like it, and you can cook it a thousand different ways. I heard a story on the radio the other day, though, that said the squid-fishing industry was growing so fast that squids were being overfished, just like salmon, and that dolphins and seals that eat squids could starve if that happens. Another environmental disaster. Now I feel guilty buying squid. I feel like I have to do a research project on my entire shopping list to find out if things are ethical to buy. I wish more foods had clear labels, like the tuna cans that say "Dolphin Safe" on them.

Squid Interest Group
Restaurant Owner

Listen, running a restaurant is hard work. Not all restaurants make it. If I don't have things on the menu that my customers want to eat, and if my food isn't reasonably priced, my restaurant won't make it either. I can't keep track of how every fish in the ocean is caught and whether it's being overharvested. A woman came in the other day and told me I shouldn't serve calamari anymore until the fishery is better regulated. I told her, ya, when it's better regulated, it'll also cost her three times as much to order! If her conscience won't let her eat calamari, fine, she doesn't have to order it, but it's not my job to make that decision for everyone who walks in here for dinner.

Squid Interest Group
Environmentalist

It's hard to believe that another fishery is on the verge of collapsing. This has happened so many times, but we never seem to learn from our past mistakes. If we put strong limits on the squid fishery now, it's still early enough that we can save it—and save all the other animals that depend on squids for food—dolphins, seals, sea lions, birds, and fish. I know the fishers are trying to make a living, but if the fishery collapses they'll really have trouble making a living. Everyone wants to "study the situation" before we make any decision. But by the time we're finished studying, the squids will all be gone. For once, let's take action before we have a disaster—not after. I think we should have an immediate halt: not let any more boats join the fishery until we know how much can be safely caught each year.

Squid Interest Group
Biologist

It seems like the fishers and the companies they sell to are only interested in making money...and that the environmentalists will never be satisfied until everyone is a vegetarian...and that the public is mostly unaware that the issues even exist...and that government agencies are just trying to keep everyone happy without taking a strong stand on anything. It seems no one's interested in the facts. If I just had the funds to put together a team to study squid biology for a couple of years, I could probably figure out a way to allow the fishery to continue without overfishing it. That doesn't seem likely, though. I hate to say "I told you so," but it wasn't hard to predict that the cod, abalone, salmon, and halibut fisheries would collapse. I'm afraid squids are going the same way. The trouble is, if squids are overfished, it's hard to predict what'll happen to the ecosystem—there are so many animals that depend on squids for prey, it could throw the entire ecosystem out of balance.

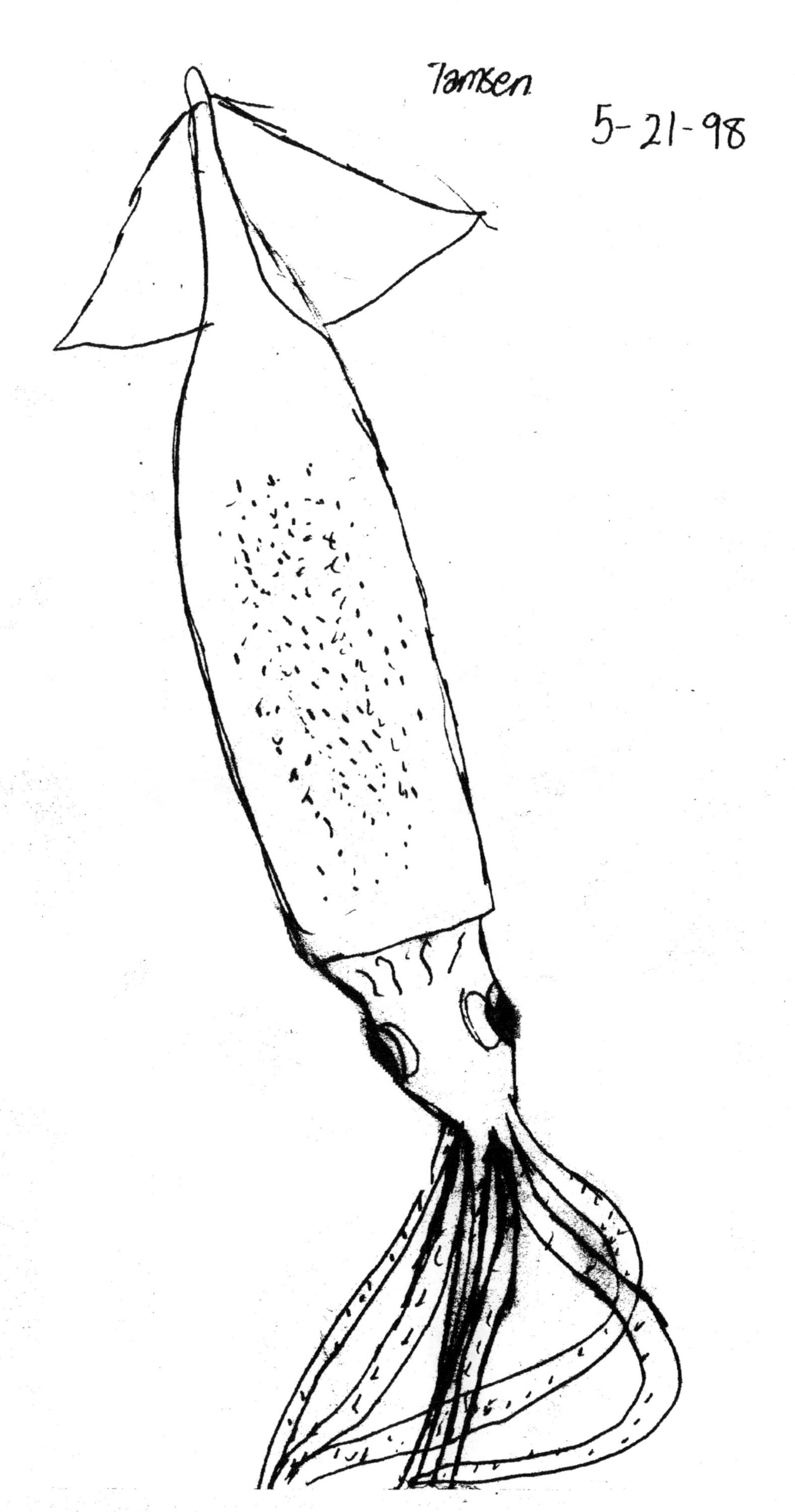

Brown
Black
Tan
orange
[illegible]
[illegible]

Activity 3: What's the Catch?

Overview

The ocean and the resources we take from it have always seemed limitless. In Apples and Oceans, students learned that this is not the case. Nearly all our fishing is concentrated in a tiny fraction of the ocean. Can it continue to feed the ever-increasing world population? For the first time in history, fisheries of the world are declining. Too many boats are taking too many fish. Too many people are adversely affecting sensitive coastal habitats necessary for the reproduction of commercially important fish and shellfish. The ocean is no longer considered an inexhaustible source of food for people; its resources must be much more efficiently managed. It's been said that pork producers, when they harvest their hogs, "use everything but the squeal"; fishing fleets must move in the direction of using "everything but the splash."

The National Science Education Standards *emphasize the importance of developing in students a more conceptual understanding of global environmental problems, and ways in which they can contribute to solutions. The World Fisheries Conference in this activity provides students an opportunity to formulate and share solutions of their own, and to listen to others who may have different ideas.*

In this activity students work cooperatively in small groups to learn about five fisheries that are among the most overexploited ocean resources across the planet. They learn that most large commercial fisheries in the open ocean flourish where the interaction of currents and sunlight provide a productive environment. Students discover that most of the ocean fisheries (13 out of the major 15) in the world are severely threatened due to overfishing or habitat loss, and that many result in additional ecological damage to the ocean because of associated ***bycatch*** (the incidental taking of species other than those targeted by the fishery). The class learns that personal choices about what we buy and eat, and our decisions about public policy, can help create fisheries that are sustainable—as long as those decisions are based on objective information about ocean systems. In the culminating sessions, the students present their findings to the rest of the class in a simulated World Fisheries Conference.

In Session 1, Seafood Smorgasbord, students get "**Into** the Activity" by sampling various seafood, and then discuss what they know about fishing and fisheries in a Thought Swap. They go "**Through** the Activity" with discussion of a reading assignment on how real people have made a difference in improving one fishery (tuna). In Session 2, Fishery Experts, cooperative groups become "panels of experts" on each of five different fisheries, and every student completes a poster with text, graphs, and/or pictures for group presentation in the next session. In

It's important to note that the fisheries data presented in this activity were representative at the time we went to print. These data (and fishery trends in general) fluctuate—even change dramatically—as scientists and fishers report new findings almost by the day. The statistics in What's the Catch? are a snapshot of the fisheries at one moment in time; they're representative, not conclusive. Your students will no doubt come up with different information while doing research—especially if they use the World Wide Web. Their findings may often show even greater decline in fish populations—testament to the urgent need for action to protect our world fisheries.

Session 3, the fishery expert groups present their findings in a poster session at the World Fisheries Conference, as each student describes her assignment and poster. In Session 4, the class makes recommendations to the nations of the world to help manage fisheries in sustainable ways, and the class compiles the conference proceedings. Students then use the **Think, Pair, Share** activity structure to clarify their own personal decisions and choices. A number of "Going Further" activities are suggested for going "**Beyond** the Activity."

What You Need

For the class:

- ❑ 25 sheets of chart paper (approximately 27" x 34")
- ❑ 1 set of colored markers (wide-tipped, water-based)
- ❑ 1 box of toothpicks
- ❑ paper towels or napkins
- ❑ 5 or more different seafood samples, enough for each student to have a small taste of each (see "Getting Ready," page 91)
- ❑ 1 picture of each of the live animals from which the seafood came
- ❑ a 3-hole punch
- ❑ 3 brads for class book, *World Fisheries Conference Proceedings*
- ❑ a roll of masking tape
- ❑ at least 5 small paper clips
- ❑ *(optional)* 1 globe or world map
- ❑ *(optional)* 4–5 marine-life reference books (see "Resources," page 163)
- ❑ *(optional)* 1 world atlas
- ❑ *(optional)* book(s) of *haiku* poems

For each of five Fishery Expert groups:

- ❑ 1 set of 6 Fishery Assignments (masters on pages 141–147; see "Getting Ready")
- ❑ 1 copy of the World Map Handout (master on page 148; see "Getting Ready")
- ❑ 1 set of 6 Fishery Information Cards (masters on pages 109–140; see "Getting Ready")
- ❑ 1 Overview of the World's Ocean Fisheries handout (master on page 108), copied on colored paper
- ❑ 2–3 sets of colored markers (wide-tipped; including red and blue)
- ❑ 4 sheets of grid paper for graphing assignments

For each student:

❑ 1 Dolphins and Your Tuna-Fish Sandwich handout (master on pages 105–107)
❑ 1 sheet of 8 ½" x 11" lined paper
❑ 1 sheet of 10" x 14" or 11" x 17" blank paper, for posters

Getting Ready

1. Make one copy of **each of the six different Fishery Information Cards (pages 109–140) for each of the five different fisheries** (you'll have a total of 32 sheets of paper):

- **What's Known about It?** for Swordfish, Tuna, Squid, Pollock, and Shrimp;
- **What's for Dinner?** for Swordfish, Tuna, Squid, Pollock, and Shrimp;
- **Where in the World?** for Swordfish, Tuna, Squid, Pollock, and Shrimp;
- **How Are They Caught?** for Swordfish, Tuna, Squid, Pollock, and Shrimp;
- **What's Happening with the Fishery?** for Swordfish, Tuna, Squid, Pollock, and Shrimp; and
- **What's the Big Deal?** for Swordfish, Tuna, Squid, Pollock, and Shrimp.

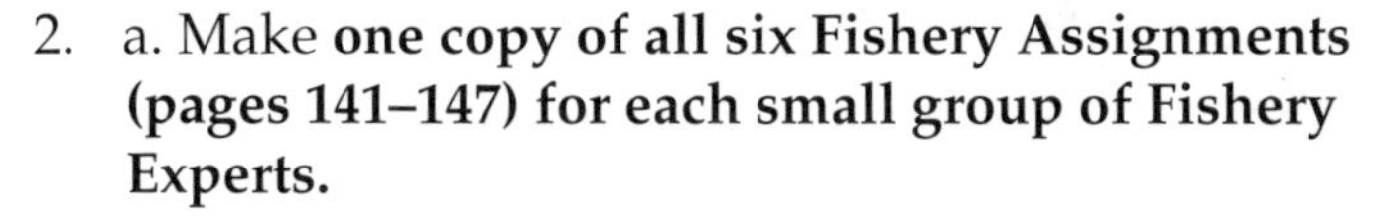

2. a. Make **one copy of all six Fishery Assignments (pages 141–147) for each small group of Fishery Experts.**

b. **Be sure to include the World Map Handout (master on page 148) with Fishery Assignment #3,"Where in the World?"** Depending on your class's knowledge of geography, you may choose to label the map or have the students doing this assignment find and label the countries themselves. (Have a globe or world map available for them to consult.)

c. Paperclip the assignments in sets of six for each group.

3. If possible, obtain five or so marine-life reference books so that each group can have access to one. **Try to ensure that each of the five fishery species is represented in each book.** Making these books available isn't a requirement but is highly recommended, so the students can do additional research on their group's fishery species—how it's caught, its predators and prey, etc.—and make guesses

Whether or not you provide extra reference material, be sure your students understand that there will be no one right response to any assignment. If the information they're looking for isn't on the provided Fishery Information Cards, students can surmise that they may have hit on a topic that requires more research—which is the whole point of "What's Known about It?"! If time permits, students can do more research on the Web at home or at school.

during the Seafood Smorgasbord. (See "Resources" on page 163 for book ideas.)

4. Decide if you'll invite people from the school (another class, the principal) and/or community (parents, conservation groups) to the World Fisheries Conference. You might have the class design a flyer advertising the event.

5. Read the Dolphins and Your Tuna-Fish Sandwich handout and the Overview of the World's Ocean Fisheries. Write key vocabulary on chart paper: ***sustainable, maximum sustainable yield (MSY), target species, bycatch (incidental take), habitat destruction, seafood,*** and ***overharvesting/overfishing.*** (See "Behind the Scenes" on page 149.)

6. Write out the Key Concepts for this activity in large, bold letters on separate strips of chart paper:

- **Most large commercial ocean fisheries flourish where the interaction of currents and sunlight provide a productive environment.**
- **Most of the ocean fisheries in the world are severely threatened due to overfishing or habitat loss, and most commercial fishing results in significant "bycatch."**
- **Personal choices about what we eat can influence public policy and the sustainability of fisheries. Scientific information should be used to help make wise choices.**

7. Just before presenting Session 1, go shopping for seafood to serve at the Seafood Smorgasbord. These products are usually available in large grocery or gourmet food stores. It's ideal (but not necessary) to have samples of the seafood caught in the five fisheries described in the activity:

- **Pollock**, which can be found in some fish sticks and prepared fish and chips, in imitation crab (sometimes sold as "krabmeat" and often found in store-bought sushi called "California rolls"), and in imitation scallops.
- **Tuna**, fresh or canned.
- **Swordfish** (this will be the hardest to find, and the most expensive).
- **Squid** (calamari), available frozen in most grocery stores.
- **Shrimp** (prawns), available frozen in most grocery stores.

You might also have samples of any of the following: tinned anchovies or sardines; real crab or scallops; canned or jarred oysters; octopus; herring (pickled or other); jellyfish; any cooked fresh or frozen ocean fish; mussels; and dried, smoked, or fresh salmon.

8. Prepare trays for each table with the following "Smorgasbord" elements, so the tasting can be set up quickly in the middle of Session 1: one ready-to-serve **unlabeled** seafood sample; toothpicks; napkins; sheets of paper for the students to draw their guesses on; pencils and colored markers; and one of the five marine-life reference books, if available.

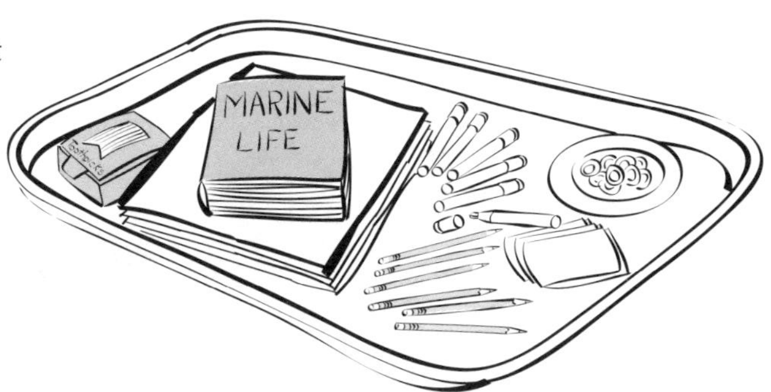

Session 1: Seafood Smorgasbord

1. Have students work in small groups to brainstorm and record lists of ocean animals that people eat. Lead a class discussion, posting their ideas on the chalkboard or chart paper.

A smorgasbord is a buffet consisting of a variety of small hot and cold offerings. The word comes from the Swedish smörgås, *meaning "open sandwich," and* bord, *meaning "table." The traditional Scandinavian smorgasbord consists of smoked and pickled fish, relishes, cheese, salads, etc., but the term is now used more liberally, for any collection of varied foodstuffs.*

2. Tell them they're going to have a seafood smorgasbord right here in the classroom, to sample some food from the ocean. As they sample, students will try to guess what each organism looked like while it was alive. Explain that each of them will be starting a piece of paper with drawings of his guesses, and taking that paper from table to table for each new seafood.

3. Set up the smorgasbord around the classroom.

4. Divide the students into groups and have each start at a different table. Letting them know you'll be setting a time limit (about five minutes per table), have them taste the food item on their table, then take a piece of paper from the stack and on it draw a picture of what they think the animal looked like when it was alive. They can flip through the marine reference book to try to figure it out. Rotate the groups after each "time's-up" (reminding students to take their drawings with them to the next table, to add to), until all students have sampled from each table and illustrated their guesses.

A small percentage of people are allergic to shellfish, and you may have vegetarians in your class who prefer not to try the samples. We encourage you to respect the needs of your students and not force students to try things they feel they shouldn't eat. At the same time, some students may just need some healthy encouragement to try something they THINK they won't like.

5. After the smorgasbord, go back to each item, one at a time, and hold it up. Have students share their guesses and briefly discuss their detective work. Why did they guess the animals they did? If no one guessed the animal correctly, show the class a picture of it and tell them its name.

Introducing Thought Swap

Note: This activity structure helps students talk and write about their related prior knowledge. It emphasizes cooperation and social skills development, and creates opportunities for students to use language in a non-threatening, highly relevant setting. Students also build on their active-listening skills by learning how to hold short, interesting discussions with a variety of partners.

1. Tell students that in Thought Swap, they'll take turns talking with different classmates. They'll need to cooperate, follow directions, and talk quietly with each of their partners.

2. Have students stand shoulder to shoulder in two parallel lines, so that each person is facing a partner.

3. Tell the students you'll be asking a question or giving them an idea to talk about with their partner (the person facing them). They'll each have about a minute to talk before "swapping" to let the other person talk. For a successful discussion, each partner should be a good listener and speak clearly when it's her turn.

You may want to pass out ocean-related pictures to encourage discussion.

Thought Swap Begins

This is how Thought Swap will go:

a) you pose a question from the list;

b) partners swap thoughts;

c) you call "time";

d) you debrief a few students;

e) students swap partners;

f) you pose a question, debrief, and so on.

1. Begin the Thought Swap by posing this two-part question for students to discuss:

- **How often do you normally eat seafood? What's your favorite kind?**

2. Walk along the two lines to help shy or reticent partners get started. When you call time, have a few students report something their partners told them.

3. Move Thought Swap along: Have one of the lines move one position to the left, so everyone is facing a new partner; the person at the end of the line walks around to the beginning of the line. Everyone can now greet his new partner. Ask the next question for the new partners to discuss:

- **What kinds of fish have you caught or have you seen someone else catch?**

4. Again, walk among the students to help the discussion along. After calling "time," have a few students describe their partners' responses.

5. Continue with the Thought Swap sequence, posing each of the questions below. After each question, debrief the students by calling on a few to describe their partners' responses.

- **Describe how you think the seafood you just tasted may have been caught.**
- **Think about the activity Apples and Oceans. Where do you think most of the fisheries of the world are located? Why?**
- **Recall the meaning of "overfishing." Can you think of an example?**
- **What can people do to make sure that enough fish will be around for future generations to eat?**

6. After calling "time" for the last question, spend some extra time discussing it with students. Write down their responses on chart paper, to be revisited after the class learns about actual fisheries.

7. Distribute the handout, Dolphins and Your Tuna-Fish Sandwich, as homework (or an in-class assignment) to be read in time for the next class. Tell students they'll be reading about what real people *have* done to try to solve problems associated with one specific fishery: tuna. Let them know they'll be studying tuna—and four other fishery species—in greater depth later in the activity. They'll have lots of opportunity to share what they learn about all those species with the rest of the class.

World Fisheries

1. Lead a class discussion about the reading assignment, Dolphins and Your Tuna-Fish Sandwich. Discuss the following questions as a whole group:

- Are you concerned about overfishing? About other environmental problems related to fisheries, such as bycatch? Why or why not?
- Do you think any of the seafood we ate in the Seafood Smorgasbord may have come from species that are overfished?
- What do you think a "scientific" approach means? An "activist" approach? Which strategy for problem solving do you think is the most effective? Is there only one "right" answer?

2. Call attention to the three major problems facing fisheries: ***overfishing***, ***habitat destruction***, and ***bycatch***. Ask students to share any proposed solutions.

3. Introduce the concept of ***maximum sustainable yield***, or ***MSY***: the total amount of fish that can be caught and still leave enough to reproduce, so that the population doesn't continually get smaller. Students will come across this term again in their fishery assignments in Session 2.

4. If you wish, paraphrase for the class some of the important information on the Overview of the World's Ocean Fisheries, before distributing it to each small group in Session 2.

In this activity we often refer to the ***maximum sustainable yield (MSY),*** *a long-standing regulatory tool for managing fish stocks. But we may not really* know *the "maximum" for any given fishery—and by dictating a quota, we risk making a mistake about the sustainable harvest. So the concept of MSY is gradually being displaced by other management ideas. One other approach uses the* ***optimum sustainable yield (OSY),*** *in which fishing limits are based not just on total populations of fish, but also on safety for the fish stock, or highest economic value, or most jobs created for fishers. Another is the* ***acceptable biological catch (ABC),*** *which takes evolutionary and biological oceanic changes into account, unrelated to the fisheries.*

Session 2: Fishery Experts

Fishery Information Cards

1. Tell students they'll now investigate five fisheries from around the world and compare how well or poorly they're managed. Divide the class into five groups. Let students know that each group will become the "expert" on a different fishery and then present its findings in the form of posters during the World Fisheries Conference. The purpose of the conference is to make recommendations for managing global fisheries.

You'll notice that several definitions and descriptions in the Fishery Information Cards ***(maximum sustainable yield****, for instance, or* ***purse seine****) are used again and again, in different sections. This is to ensure that every student has the information necessary to complete her particular assignment.*

2. Give each small group a set of six Fishery Information Cards for **one** fishery species, and six sheets of lined paper. Be sure to hand out a card to every student in every group, and call their attention to the different icon or drawing depicted on each card.

We've made every effort to keep the Fishery Information Cards clear, accurate, and accessible, and to define or explain terms and concepts that arise. Depending on your students' grade level, the information may still be challenging. You know your students best: feel free to adapt the cards as needed! We welcome your ideas for adapting for younger students.

Note: Fishery Assignment #6, "What's the Big Deal?," asks students to convey fisheries information not only with graphs, but also using either language (poetry or prose) or visual art (drawing or cartooning). You may wish to give this Fishery Assignment and its corresponding Fishery Information Card to the student in each group you think would best express himself in (or most benefit from) one of these alternative learning styles.

3. Have students read their cards silently and write down the two or three most important or interesting points.

There are six Fishery Information Cards. If any group of students is larger than six, you may want to have some of them work on the cards (and assignments) with a partner.

4. Have each student in turn summarize and share highlights from her card with her small group. (If a student has encountered terminology that isn't explained in her card, this is the time to ask fellow group members about it—another student may have it on his. Circulate to check for understanding and to answer questions.)

Some of the assignments in this session require students to create graphs. Using graphs to represent their findings provides your students with an opportunity to apply what they know about data representation in a meaningful way. Depending on the experience of your class, you may need to review how to make various graphs, including bar graphs, circle graphs (like the one in Apples and Oceans), and line graphs. The Glossary of Graphs on page 156 makes a good warm-up.

Students from different groups with the same graphing assignment can also come together and discuss how they'll represent the information about their fishery species.

Fishery Assignments

1. After the small-group discussions, distribute the following to each group:

a. One packet of Fishery Assignments. Ask each student in the group to pick out the assignment **with the same name and icon as on his Fishery Information Card**. Each assignment directs the student to make a particular poster for his group's presentation at the World Fisheries Conference.

b. One copy of the Overview of the World's Ocean Fisheries, copied on colored paper, to be used as a table reference while the students work.

c. Grid paper, for those with graphing assignments.

d. A marine-life reference book, if available.

2. Direct students to the art supplies and paper available for use in making the posters. Remind them that some of the assignments include a form of graph as well as other representation. The posters from all the groups will be bound together as the *World Fisheries Conference Proceedings.*

3. Students may need to complete their posters as homework. Remind them that in the next session they'll present their posters during the World Fisheries Conference.

Several of these assignments require manipulation of considerable data. These are exactly the kinds of data real fishery managers use to find trends, or patterns, that help them protect the fisheries. Unless we can determine trends, we can't really know what's happening with the fisheries. Encourage students to be detectives and to tackle their charts, tables, and statistics with enthusiasm!

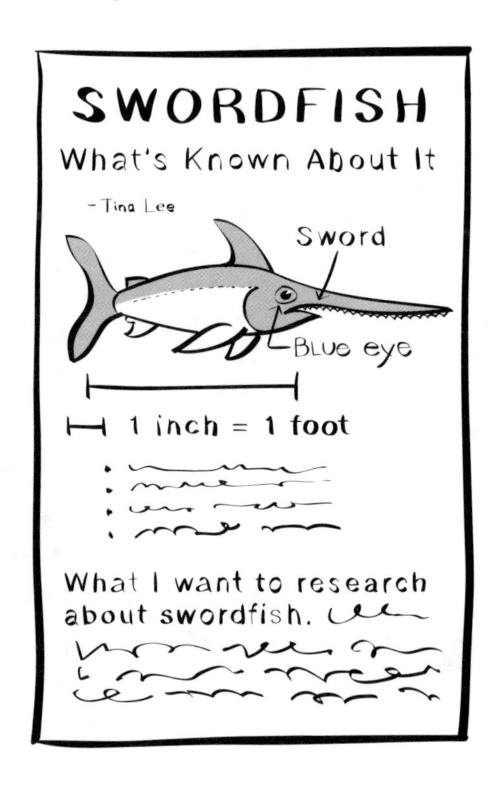

"Today a neighbor came by bringing us some fresh fish he had caught on his most recent boat trip. As we thanked him, he spread his arms, saying, 'They are not from me; they only come through me.' At that moment I realized a similarity between fish and *haiku*."

—Jane Reichhold, from www.ahapoetry.com/haiku.htm#oceanku

Session 3: The World Fisheries Conference—Poster Presentations

1. After all students have completed their assigned posters, have each small group briefly discuss the following pre-conference questions as they relate to its fishery. Assign one member of each group to takes notes.

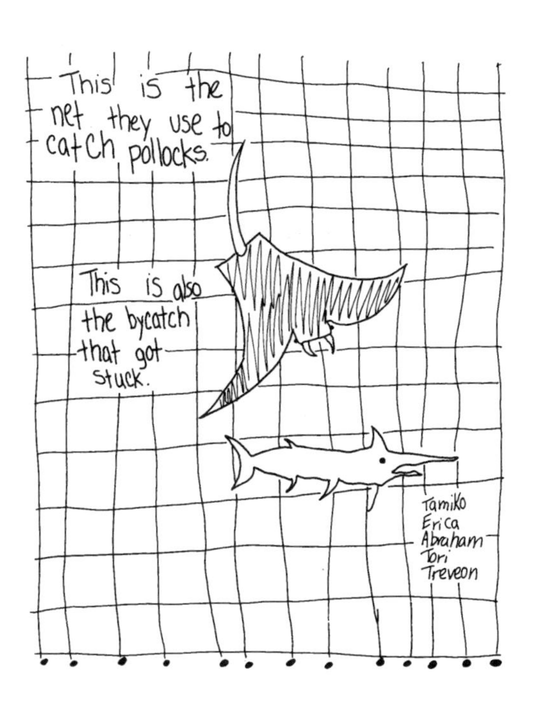

Pre-Conference Questions

- Do current fishing methods take the biology of the animal into account?
- What recommendations would your group propose to better manage the fishery?
- What actions can individual citizens like us take now to ensure this species is around in the future?

2. Have each group make a presentation to the rest of the class as the "panel of experts" on its fishery, with each member of the group presenting her poster in turn. Each group should end its presentation with a conclusion and discussion of the pre-conference questions, and ask if there are any other questions from the class.

3. After all groups have presented, lead a discussion about what all the fisheries have in common. [Almost all are located in highly productive parts of the ocean: areas of upwelling or near the poles (see sidebar), or in estuaries, where rivers meet the ocean. They're also nearly all declining, and almost all are poorly managed.]

The North and South Poles, with their 24 hours of summer sunlight and abundant nutrients in the water, teem with biological productivity.

4. Collect, three-hole punch, and compile all the posters into the *World Fisheries Conference Proceedings,* holding the book together with brads.

Session 4: The World Fisheries Conference—"Big Picture" Recommendations

1. Have the class imagine itself at the closing session of the World Fisheries Conference. The students' job is to make general recommendations to the nations of the world to help to protect the world's fisheries, to make sure

that 1) there is enough food to feed the world in the future, and 2) the biological health of the ocean is maintained.

2. Have the students work in the same fishery groups as before. Distribute to each group one 8 1/2" x 11" sheet of lined paper on which to record its recommendations. Their suggestions may be new ones, based on what they heard in Session 3 from other groups, or reflect their own poster presentations. Have one member of each group act as recorder, to write down the group's ideas. Several examples of possible recommendations follow, but it's best not to mention these until *after* the students have come up with and recorded their own ideas. You may want to post it during the discussion that follows and chart any areas of overlap.

You may wish to make a transparency of this list and have it ready to show on an overhead projector.

Possible recommendations:

- Don't catch any fish before they're old enough to reproduce.
- Reduce catch in every fishery by 10 percent.
- Study every fishery species that is being commercially harvested.
- Encourage scientists to work in partnership with fishers, who have great knowledge of the ocean and of their fishery species.
- Decrease the need for seafood by controlling the population of people.
- Develop alternate sources of protein, like soybeans, for feeding people.
- Tax seafood to discourage people from eating it.
- Create more marine reserves that prohibit fishing.
- Don't allow commercial fishing in sensitive habitats; create more international Marine Protected Areas.
- Make laws to punish excessive bycatch.
- Label seafood to show which products use fish caught in environmentally responsible ways.
- Have all nations ratify (accept and pass) the existing treaty that protects fish crossing the borders between countries.
- Stop the growth of fishing fleets.
- Spend more money to learn about what we catch, instead of just trying to catch more.
- Strengthen and enforce environmental laws, including the Clean Water Act, the Endangered Species Act, and the Sustainable Fisheries Act.
- Increase the number of federal "fishery cops," to make sure the laws are being enforced.
- Use fishing gear that doesn't destroy habitats or kill unwanted marine life.

Michelle
May 20, 1998

Tuna

America does most of the fishing. America eats most of the catch. Yes, I think El Niño will affect the size of the catch. If there isn't any tuna it effects it. It says tuna swim around, if tuna all swim away there won't be any more and it will effects it. Stop people from fishing so there will be some in the future. Put it on T.V., news, and radio ads ect.

- Discipline fishers who don't use bycatch exclusion techniques such as TEDs, BRDs, and "backing down."
- Offer assistance to fishers who follow rules that limit their catch or require new equipment.

3. List all group recommendations on chart paper or a transparency, putting a star by recommendations made by more than one group. If you post the additional possible recommendations above, the class can discuss areas of overlap with those as well. Each group can add recommendations to its own list as they're added to the class list. When you're finished, add each group's list to the class book, *World Fisheries Conference Proceedings.*

Donald
May 10, 1998

Shrimp

1. Because people want it for food, the population is doing bad because they're all dying off. They also make laws to reproduce shrimp. If I were to do something about what's happening, I would limit it to how many shrimp you can catch.

Think, Pair, Share: My Values and Behavior

1. Have the students recall how to do a Think, Pair, Share from Activities 1 and 2. Tell them they'll be asked to think about what they've learned about global fisheries, and to make some decisions about their own personal choices.

2. Distribute a sheet of lined paper to each student. Have students **Think** about, and list on their papers, the names of all five fishery species from the World Fisheries Conference. Next to each species, have them write whether or not they'll choose to eat that seafood in the future. If so, how often and why? If not, why not? (Ask students to clarify which fish they chose not to eat because of personal taste preferences, rather than reasons related to the fishery.)

3. For **Pair** and **Share**, have students discuss and compare their answers with a partner.

4. Continue the sharing with a class discussion; you may want to tally the results by species into a bar graph on chart paper or on the chalkboard. Some students will probably be surprised by other students' personal decisions. For example, some students may decide that they'll never eat shrimp—or that they'll only eat it if they know for sure how it was harvested. Other students may decide they aren't going to worry about shrimp and will eat it whenever they want to, but perhaps decide never to eat swordfish. Discuss whether students believe their personal choices make a difference. Why or why not?

It's important to stress to students that there are no "right" or "wrong" responses in this discussion. Part of the exercise is learning to accept differing opinions without criticism, and it's likely students will feel freer to express their personal decisions if the discussion is nonjudgmental.

5. Distribute blank paper and colored markers to each student. Students can now design individual **mascots** (symbols or cartoons) or **slogans** (sayings) they think would persuade others to support their viewpoint. Remind them of how powerful these images and sayings can be. (Think of Smoky the Bear for preventing forest fires, or these slogans: "Don't Teach Your Trash to Swim," which refers to keeping litter out of the ocean; "Don't Be a Litterbug"; or "Mind if I smoke? Care if I die?" and "Kick the Habit, not the Bucket," referring to the smoking habit.)

6. Display the completed posters around the room. Have half the students stand by their posters while the other half mingles and asks questions, then have them switch places.

7. In closing the activity, first hold up the Key Concepts from What's the Catch?, one at a time, and have one or more students read them aloud. Briefly discuss how these statements review the activity's important ideas.

- **Most large commercial ocean fisheries flourish where the interaction of currents and sunlight provide a productive environment.**

- **Most of the ocean fisheries in the world are severely threatened due to overfishing or habitat loss, and most commercial fishing results in significant "bycatch."**

- **Personal choices about what we eat can influence public policy and the sustainability of fisheries. Scientific information should be used to help make wise choices.**

Finally, review this Key Concept from Activity 1, Apples and Oceans, and discuss it in light of what students have learned about fisheries:

- **Most of our planet is covered by ocean, but only a small fraction of the ocean supports large concentrations of life.**

Post all the concepts on the wall for students to revisit.

Going Further

1. Consider taking out a one-year class membership to Earth Island Institute, which includes a subscription to the *Earth Island Journal* and periodic updates on such campaigns as the International Marine Mammal Project and the Pinniped Fisheries Project (pinnipeds are seals and sea lions). Visit Web sites concerned with conservation and fisheries, such as the Sea Turtle Restoration Project, and those representing different points of view, such as Greenpeace, the World Wildlife Fund, and the Tuna Association. For consumer guidelines in making responsible seafood choices, see the Web sites of the Monterey Bay Aquarium or the National Audubon Society Living Oceans Program. See "Resources" on page 163 for more information.

2. Ask professional fishers to come and talk to your class. Ask them to describe their profession, and to discuss such things as how the fish are caught and what problems and issues they perceive in their fishery. See if they can bring in a sample of their catch and some of their gear.

3. Take the class to visit a fish market or seafood restaurant and ask where the fish were caught. Are any of them local species? How far away were they caught? Were they raised using aquaculture (underwater "farming")? Ask your class if anyone eats these species. Back in the classroom, do some research on those species and locate their habitats on the map. Are they well-managed or overexploited fisheries?

4. Have a "Sustainable-Fish Feed" for parents or other classes, serving only fish and shellfish species caught in the most up-to-date, ecologically sustainable manner available.

5. Take a field trip to some local fishing docks to watch the haul come in or watch recreational fishers. If you're inland, visit a lake, river, or reservoir where sport fishing occurs. Talk to the fishers about what they use as bait and where the bait comes from. Do they eat their catch or throw it back? Back in the classroom, do some research about the species they were catching.

6. Visit Chinese or other Asian markets to see freshwater and marine organisms from around the world. Do some research back in the classroom to find out all you can about

these species. Are they a farmed species? Are they overfished? How are they caught?

7. Have the class draw pictures of the marine sanctuaries and estuarine reserves around the United States. (See "Resources" for posters and information.) What sorts of activities are allowed in sanctuaries, and what are not? What about reserves? How might these protected areas be used in managing fish populations?

The book Kid Heroes of the Environment *(The EarthWorks Group, 1991) has great stories about kids who've taken action to make things better for the Earth. (See "Resources.")*

8. Have cooperative student groups write position papers describing their opinions about specific fish and fisheries, based on class discussion and global fisheries recommendations. Remind them that the tuna boycott that ended up protecting dolphins and giving consumers the option of buying "Dolphin Safe" tuna in the 1990s was largely inspired by committed students! These position papers can also be bound into the *World Fisheries Conference Proceedings.*

9. Have students write letters to the Legislature supporting their positions on coastal and ocean laws and enforcement. They can include information or pictures from the World Fisheries Conference to support their views. (At www.congress.org/congressorg/dbq/officials, students can plug in their zip code or state and instantly get the contact information and political profile of their representative(s) in Congress and the Senate. The site also lists the committees on which these officials serve; start with "Environment," and look for the "Fisheries" subcommittee under that.)

10. Have students do research on other environmental issues, such as logging, oil drilling and transport, and whaling. (Note that whaling is considered a "fishery" in its own right.)

Dolphins and Your Tuna-Fish Sandwich

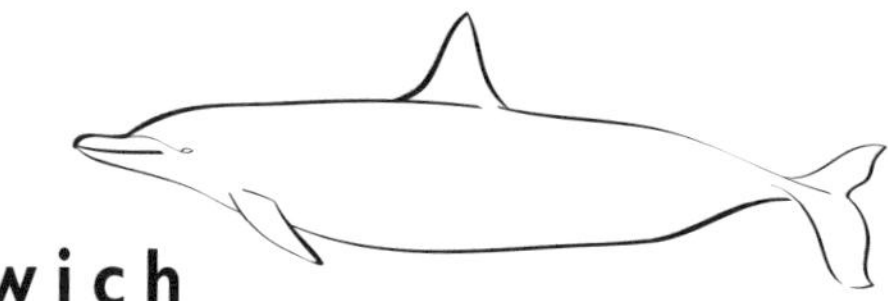

When you take a bite out of a tuna-fish sandwich, do you ever think about where that tuna came from or how it was caught? Tuna is one of the most common seafoods eaten in the United States—and the methods for catching it have created quite a stir over the last 30 years. Tuna along the west coasts of North and South America often swim under schools of dolphins. Tuna-fishing boats speed around the dolphins, encircling them in huge nets (a process called "setting" on the dolphins). Then the nets are pulled in, full of tuna...and full of dolphins. In the late 1970s, about one dolphin died for every ton of tuna caught.

There have been many efforts by people to solve this problem. There are many strategies for working to protect the environment and ensuring that people use natural resources (like tuna) in sustainable ways. Read the short stories below and think about how you would solve a tough problem like the tuna/dolphin controversy!

The Ken Norris Story

In the 1960s and 1970s, a marine biologist and dolphin expert named Ken Norris learned that hundreds of thousands of dolphins were being killed each year in tuna nets. For every 550 tons of tuna caught (to use an example you'll see again in a moment) nearly 600 dolphins were killed—and that could happen in just a few "sets" of the net! He decided to use all his skills as a scientist to find ways of catching tuna that would not kill dolphins. He spent time onboard the tuna boats, observing how they operated. He swam inside the tuna nets to see how the dolphins behaved when they were trapped. He talked with and became friends with many of the fishers to learn everything he could about their jobs and why they did things the way they did. He realized that the fishers didn't *want* to kill dolphins, but they didn't know any other way to catch tuna, which was how they made their living.

He discovered that by using more expensive, finer-mesh nets (nets with smaller holes), and by driving the fishing boats a certain way, most of the dolphins could be saved. After the dolphins had been encircled and most of the net had been hauled onto the stern (back) of the ship, if the captain quickly reversed the ship's engines it could drag the edge of the net down under the water

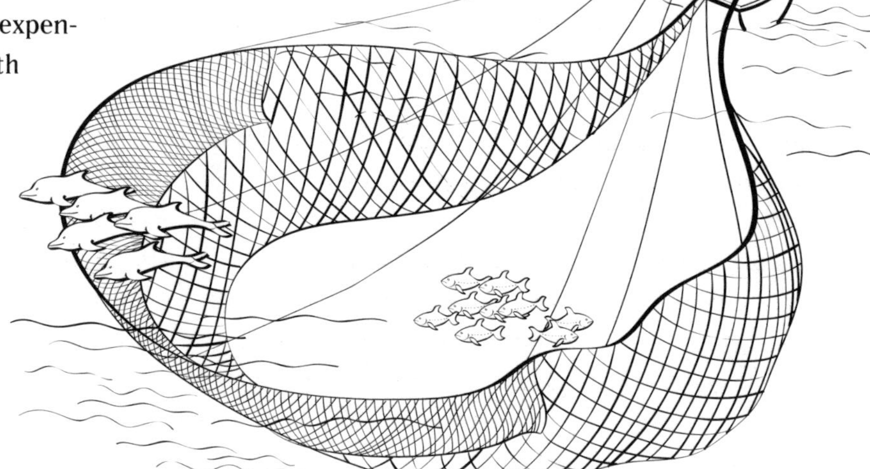

and the dolphins could easily slip out over the top of the net. He called this "backing down." He convinced one captain to try it out—and it worked! The ship caught 550 tons of tuna (like the example in the last paragraph) but killed only 11 dolphins—nearly 50 times fewer than with the old method. And Ken thought that with more practice, they could get it down to just one or two. Ken took his discovery to the United States Congress and worked hard to pass laws to force all tuna fishers in the U.S. to use better nets and employ this new fishing method. Ken helped write new laws and the rules to enforce them. He had the idea that there should be an official "observer" on every tuna boat, to make sure dolphins weren't being killed in their nets—and that if too many dolphins were killed, the fishers should have to stop fishing until the next year. Within a few years, the number of dolphins being killed decreased dramatically. Many people thought Ken was a real-life hero—he had made observations, talked with people on both sides of the issue, and quietly found a solution that was fair to both the fishers and the dolphins. It allowed the tuna fishers to keep fishing, but forced them to make extra efforts to protect dolphins.

The Sam LaBudde Story

With the success of Ken Norris, many people thought the tuna/dolphin controversy was over. But in the early 1980s, stories started to appear again about dolphins being killed in tuna nets. What was happening? Well, foreign boats had an advantage over U.S. boats, since they could use quicker, cheaper fishing methods that killed more dolphins. Many U.S. tuna boats registered in other countries so they wouldn't have to abide by U.S. laws. At the same time, there were rumors that on some U.S. boats the fishers locked the observers in their cabins for the entire trip, or threatened to hurt them if they reported what they actually saw. Few people wanted to be observers, and many boats sailed without one. At the same time, to help the U.S. fishing industry compete with foreign countries, the president of the United States relaxed enforcement of the fishery laws Ken Norris and others had worked so hard for. More dolphins than ever were dying in the nets.

That's when Sam LaBudde (pronounced "LaBuddy"), a young environmental activist, decided to take matters into his own hands. Sam got a job onboard a foreign tuna boat to secretly record what was really happening when the boats fished. Many people warned Sam that if the fishers found out what he was trying to do, he could be in serious danger. He took a video camera with him onto the boat and told the crew that he was just taking "home videos" to show his family what it was like to live at sea and to be a fisher. In fact, this is what Sam's camera recorded: In a single set (encirclement), two hundred dolphins were trapped in the nets. The captain did *not* use the "backing down" procedure, and every dolphin died a terrible death. Many were crushed as the nets were pulled in by electric winches through large metal pulleys. Many drowned. It was a catastrophe. Sam left the boat as soon as he could and released his videotape to television stations all over the country. The video he shot on that tuna boat created a small revolution! A U.S. fisher soon stepped forward with a similar tape from a U.S. boat. After seeing the tapes, environmental groups in San Francisco sued the federal government and received a court order to place observers once again on every U.S. tuna boat. Environmentalists demanded that big tuna companies like StarKist and Bumblebee stop buying tuna from boats that killed dolphins. But still, the killing went on.

Students Step In

In 1989 and 1990, after seeing Sam's video, students across the United States got together and decided to do more to stop the killing of dolphins. They organized education campaigns, wrote articles in school newspapers, and sent huge waves of letters and drawings to the tuna companies and to government officials. They pressured school cafeterias and restaurants to stop serving tuna. One group of 200 middle school students went to Washington and met with U.S. senators to explain how tuna fishing methods led to the killing of dolphins. Young people took on the biggest tuna company in the world, StarKist (which is owned by H.J. Heinz Co.); they sent boxes and boxes of letters to its chairperson. Finally, on April 12, 1990, all the hard work of students and adults paid off. StarKist announced it would no longer buy tuna caught by setting on dolphins, and began selling tuna in cans labeled "Dolphin Safe." Every other U.S. tuna company followed StarKist's example, and once again the number of dolphins being killed began to drop.

Your Turn!

This story isn't over. Despite the efforts of Ken Norris, Sam LaBudde, dedicated environmentalists, concerned government officials, and thousands of determined students, dolphins are still dying in tuna nets. There are no easy solutions that protect both the dolphins AND the fishers' right to make a living fishing. Fishing boats registered in other countries can still set their nets on dolphins—even in U.S. waters—as long as they sell their tuna to countries outside the United States. In fact, U.S. tuna fishers can set on dolphins too, if they sell their catch outside the United States. Even the "backing down" method of saving dolphins is now controversial! To understand the latest challenges and problems with the tuna fishery, search for information on the Internet, if you have access to it. (Try searching for "tuna/dolphin.")

The tuna fishery isn't the only fishery with problems. About 70 percent of all the world's fisheries are being fished to capacity (meaning the population is barely able to sustain itself), or are overfished (meaning the population is decreasing). Here's a terrible statistic: One-third of the worldwide ocean catch of 93 million tons is wasted—thrown back into the sea, dead or dying. These wasted marine animals (like the dolphins described above), caught along with the intended (target) species, are called "bycatch." Bycatch is a very serious environmental problem.

What do you think should happen next? If you got involved, what strategy do you think would be the most effective? Which would you take: the approach of a scientist like Ken, or of an activist like Sam? Could you persuade your fellow students to get involved for real? As you continue with "What's the Catch" in class, remember that it's more than a science activity. This isn't an imaginary problem. While you've been reading this page, thousands of tons more tuna have been caught and thousands more dolphins have died. Other fish and marine mammals are in danger from current fishing methods too, and many ocean habitats are being destroyed.

Will *your* ideas help to find the next solutions?

Overview of the World's Ocean Fisheries

- Most of the world's ocean fisheries are not sustainable. This means that more fish are being caught each year than can reproduce, and the population just gets smaller and smaller.

- 70 percent of the world's fisheries are fully fished (the population is barely able to sustain itself) or overfished (the population is decreasing). Thirteen out of the 15 major ocean fishing areas are now overfished. Four areas have actually been "fished out" altogether. Some parts of the ocean have been closed to all fishing, in hopes the marine population will recover.

- Fish is an essential source of food for people all over the world. Nearly one billion people in Asia alone depend on seafood as a major source of protein.

- World population growth is the biggest reason for the increased demand for fish. Ocean fisheries can't keep up with the growing demand created by overpopulation.

- About one-third of the worldwide ocean catch of 93 million tons is wasted! This 30 million tons of *bycatch* (incidental take) is thrown back into the sea, dead or dying, because it's not what the fishers were fishing for.

- One-third of all the intended ocean catch is turned into animal feed, including feed for "farmed" marine species such as shrimp.

- Nearly every individual fishery in the world is currently in decline.

SWORDFISH

What's Known about It?

Their "Swords"

Swordfish are found worldwide both in temperate seas (which never get very hot or very cold) and in the tropics (where seas are always warm). Their long bill (which gave them their name, "sword" fish) is a flattened extension of the upper jaw. It helps them swim fast by cutting through the water. The bills are also serrated (notched, like a saw or a bread knife), and are used in feeding—slashing through shoals of fish and sometimes even stabbing the prey!

How They Feed

Swordfish are very muscular, and have large, bright blue eyes. They can chase their prey in deep water with little light, and it's thought that they hunt for food at night. Amazingly, swordfish aren't cold-blooded, as are almost all other fish, but *warm*-blooded—which means they can keep their body temperatures warmer than the surrounding ocean. (Tuna and some sharks are the same way.) Their rapid metabolism allows them to swim faster and longer than cold-blooded fish, and to hunt at greater, colder depths than many other predators. It's a great advantage in the open ocean.

How Long They Live

Female swordfish live about 25 years; they grow larger than males, but we don't know exactly why. (Is it because they grow faster or live longer? Some other reason?)

Reproduction

Swordfish can begin ***spawning*** (laying eggs) at about 5 years old, when they're about 75 pounds and 3 ½ feet long. A female produces up to six million eggs per season. We know very little about swordfish breeding or migration patterns, but some swordfish have recently been tagged using sonic (sound-emitting) tags so they can be monitored and followed for days, both day and night.

What They Look Like

Like squids and many other marine organisms, swordfish use ***countershade coloration*** to be less visible in the water from both above and below; they're bronze-colored on the back and silver below, with black belly fins and no pelvic fins. Swordfish get about 15 feet long (measured from the eye to the fork of the tail), and the females can weigh 1,000 pounds or more.

Where They Hang Out

Swordfish are usually solitary fish that live between the surface and 400 feet below—though they've been sighted at 2,000 feet down!

Other Swordfishy Facts

- The scientific name for swordfish is *Xiphius gladias*, from the Greek and Latin words for "sword."
- Swordfish have a reputation for unprovoked aggression, occasionally attacking boats by ramming them with their bill. The submersible (underwater) research vessel *Alvin* was once actually rammed by a swordfish—and as the fish was unable to get loose, the crew had swordfish for dinner!

SWORDFISH

What's for Dinner?

What They Eat

Swordfish feed in every layer of the ocean, from the surface to several hundred feet deep, and will eat just about anything—including tuna, sardine, anchovy, mackerel, herring, hake, smaller swordfish, krill, pelagic red crab, and squids. Swordfish prey are often found with stab or slash marks on their bodies, which tells scientists that the bill ("sword") is sometimes used to kill or injure prey. The stomach contents of one swordfish revealed 50 pounds of prey, including 27 pounds of various oceanic fish, 12 pounds of crustaceans (krill and crabs), and 11 pounds of squid.

What Eats Them

Tuna, marlin, blue and mako sharks, and sailfish eat small swordfish. Sperm whales, killer whales (orcas), and large sharks such as great whites eat large swordfish. But humans are the major predator of this species. It's against the law in the U.S. to catch small swordfish, which are not yet old enough to reproduce—but many are caught by mistake and thrown back into the sea, dead or dying. Most of these catches are never reported.

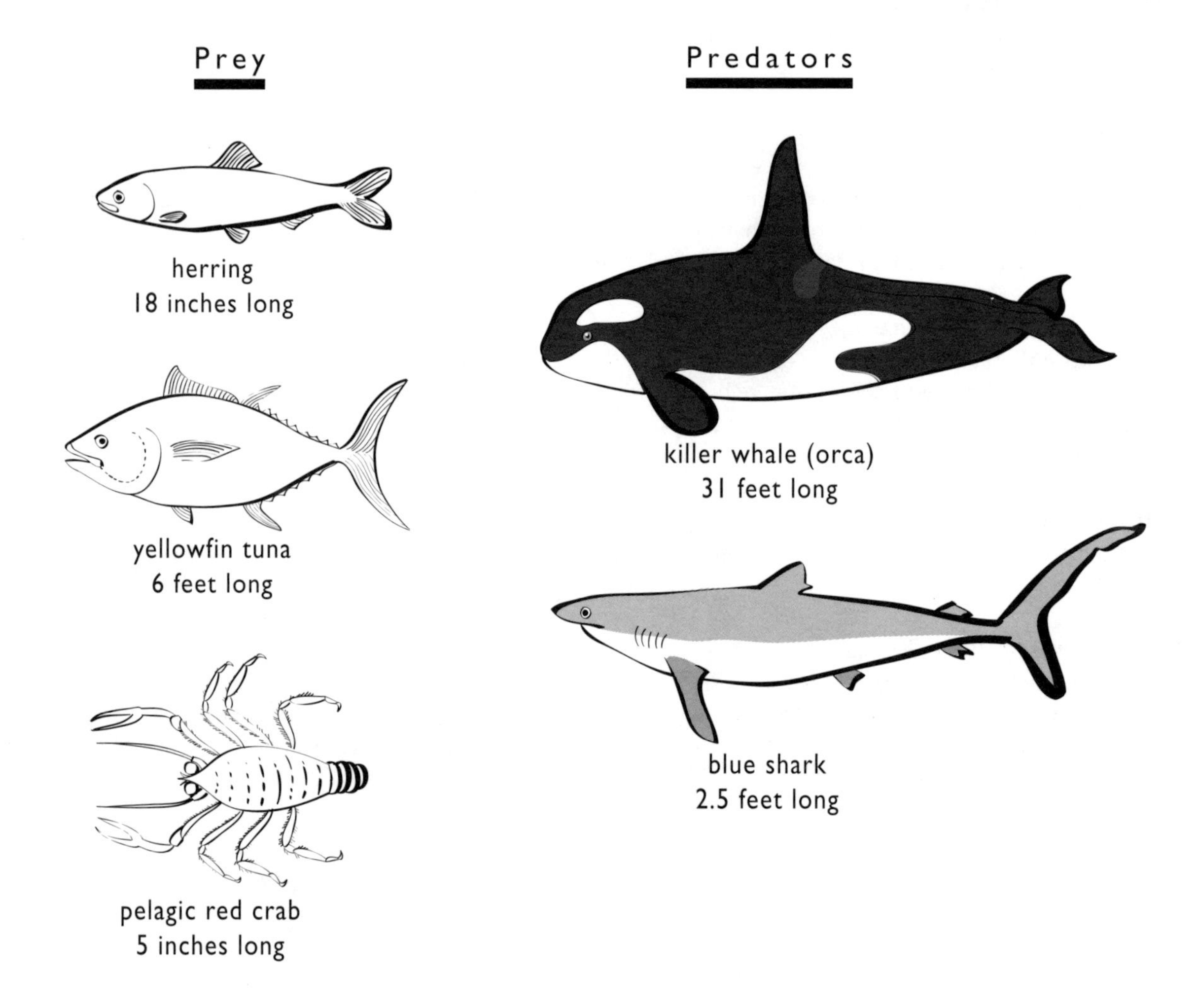

SWORDFISH

Where in the World?

Where They're Caught

Swordfish are found throughout the world both in temperate seas (which never get very hot or very cold) and in the tropics (where seas are always warm), and are fished just about everywhere they occur. The major fisheries are located in the northern and central Pacific, from just below Russia's Kamchatka Peninsula across the entire width of the Pacific Rim and as far south as the tropic of Cancer. In the southern Pacific and Indian ocean basins, they're fished off the coast of Chile, off northern New Zealand, off the east and west coasts of Australia, and east of Madagascar. In the Atlantic they're fished throughout the Caribbean and the Gulf of Mexico, off the east coast of North America across to Europe, and south across the entire southern Atlantic between Africa and South America.

Who Catches Them

Japanese boats bring in most of the Pacific swordfish, and swordfish is now the second biggest U.S. fishery in the Pacific. Brazil, Morocco, Namibia, Portugal, South Africa, Uruguay, and Venezuela have significant swordfish fisheries in the Pacific. Taiwan, Korea, France, and Brazil take large amounts of swordfish in the Pacific as ***bycatch*** (unintentional or "incidental" catch) while fishing for other species. The Japanese tuna longline fishery in the Atlantic also takes a lot of swordfish as bycatch. It's very difficult to get an accurate estimate of the swordfish catch worldwide, because much of it comes from unregulated fishing and is unreported.

The following table shows in metric tons (mt) the landings (size of the catch) for countries taking the most **Atlantic swordfish** in 1999. (Total for 1999 was 10,754 metric tons.)

Country	Swordfish Landings (in metric tons)
Spain	3,993
Canada	1,119
Japan	1,525
United States of America	2,906
Portugal	777
Taiwan	285

The following table shows the amount of fish and shellfish **of all kinds** eaten per year, per person (per capita) in countries consuming the most and least fish per person.

Country	Fish Consumption per Capita (in pounds)
Japan	152
China	53
United States of America	46
India	10
Indonesia	40
Cambodia	20
Spain	90

Fishery Information Card

SWORDFISH

How Are They Caught?

The Fishing Method

Almost all swordfish are caught by ***longline*** fishing at night, when they feed at the surface. Each longline has up to 3,000 shorter lines attached to it, each with a baited hook at the end. A set of longline gear from a single 130-foot boat can stretch up to 80 miles across the ocean! Longline vessels travel far offshore, fishing for several months at a time, day and night.

What Else Gets Caught

Longline fishing accidentally kills thousands of other, untargeted animals (called ***bycatch***), including bluefin tuna and the sea turtles and albatross (large seabirds) that are attracted to the bait and get caught before the hooks can sink below the surface.

Managing the Fishery

The faster the hooks sink, the less bycatch of birds and turtles there is—so North Pacific longline fishers must now used weighted hooks They're also required to use fewer lights on the boat when they fish at night, so that seabirds are less attracted to the bait. And they have to hang lines above the baited hooks to scare the birds away.

A Good Management Example: Hawaii

Rapid growth of the swordfish fishing fleet has also hurt some endangered species, including the Hawaiian monk seal. No longline fishing is allowed around the Hawaiian Islands, which are home to monk seals, three species of albatross, and several types of sea turtles. Commercial fishing is being moved to areas where bycatch will be lower. Protected areas are closed to swordfish fishing at certain times of year. All boats around Hawaii are monitored every hour to make sure they're staying out of protected areas.

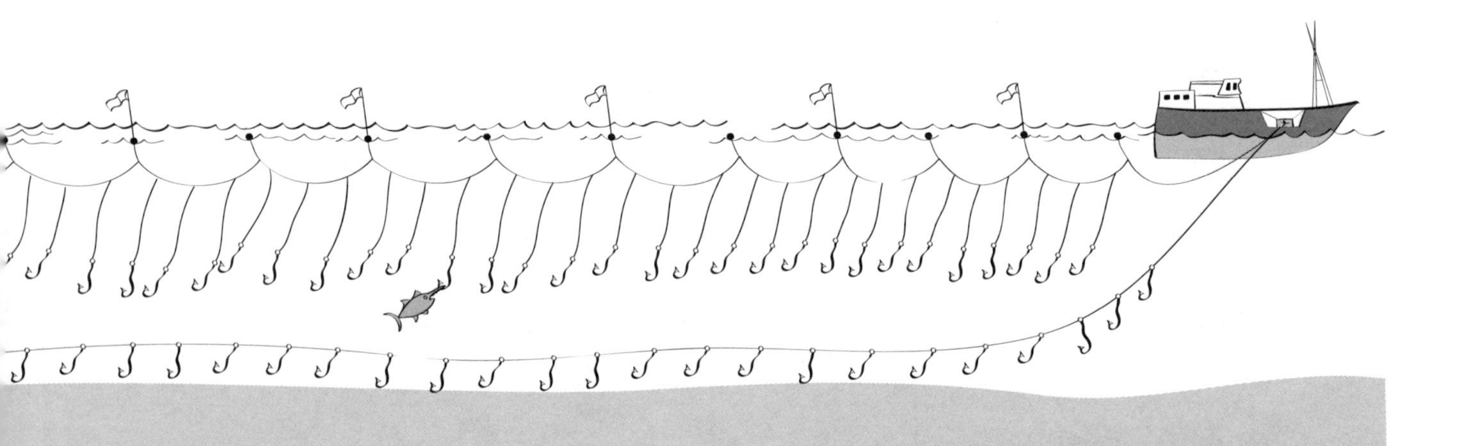

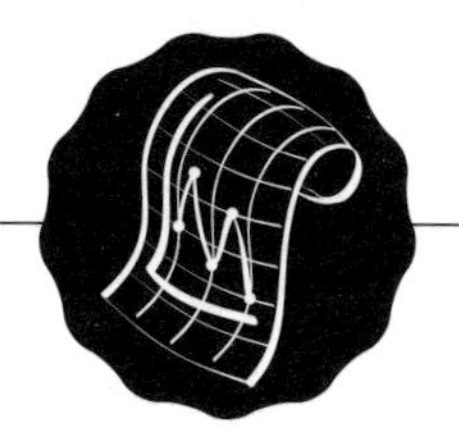

SWORDFISH

What's Happening with the Fishery?

Disappearing Swordfish

Thick, meaty swordfish steak was once a popular menu item in expensive restaurants. That's changed, though, as swordfish continue to disappear due to overfishing. Many swordfish were once caught off the Atlantic coast of the U.S., but they're now considered overfished, and many restaurants no longer serve them (see note below). In 1982, swordfish made up 60 percent of all the open-ocean fish caught by U.S. fishers in the Atlantic, but the swordfish population has almost disappeared because of continued overfishing.

Note: In 1998, a huge campaign evolved to persuade chefs and other fish providers around the U.S. not to sell North Atlantic swordfish in their restaurants and stores. The "Give Swordfish a Break" campaign had the same kind of effect as the "Dolphin-Safe Tuna" efforts you read about in Dolphins and Your Tuna-Fish Sandwich; it gave an at-risk species a needed break, and heightened consumer awareness about the state of the fishery.

Setting Limits

The amount of fishing that can occur while still leaving enough animals to reproduce is called the ***maximum sustainable yield (MSY)***. In 1999, the international group responsible for swordfish management estimated that the yearly swordfish catch—the MSY—must not exceed 10,000 metric tons. This would allow the swordfish population to grow despite fishing pressure.

Regulating the Fishery

Establishing a maximum sustainable yield in 1999 was part of a plan by the swordfish management group to rebuild North Atlantic swordfish populations within 10 years. Conservationists, the U.S. government, and the U.S. fishing industry pushed hard to *immediately* reduce the number of fish that could be caught, but the European Union and Japan did not agree. So the compromise measure that was agreed on by everyone included a *gradual* catch reduction over three years—not enough to rebuild the population within 10 years, but better than no reduction at all.

Scientists are also very concerned about the huge numbers of small, immature swordfish caught unintentionally, thrown back dead, and often not counted. In 2000, the U.S. government decided to make swordfish nursery grounds—where young swordfish develop—"off limits" (closed) during the breeding season, hoping to reduce the number of young swordfish caught.

Enforcing the Law

Rebuilding the fishery is a big job. It all depends on monitoring and enforcing the nursery closures and other regulations. Unfortunately, although the U.S. uses observers to monitor and count the discarded ***bycatch*** (dead fish thrown back into the sea), some other countries do not. Spain—which catches the most swordfish in the North Atlantic—has no observers on board...and it reported "zero" discards in 1999. Was this true?

(continued→)

The Pacific Fishery

When the Atlantic fishery for swordfish collapsed, the fishing fleet moved to the Pacific. (The chart below shows how too much swordfishing over the years finally hurt the population in the Atlantic.) Japan and the U.S. bring in most of the Pacific swordfish, and swordfish is now the second biggest U.S. fishery there. But the total catch in the Pacific has declined each year since 1993 due to overfishing. All countries fishing for swordfish need to cooperate to protect the population. In 1995, the United Nations produced a treaty to protect fish, like swordfish, that cross the coastal zones of two or more countries or migrate through international waters. The treaty will only take effect, though, once 45 nations have agreed to it.

The table below shows in metric tons (mt) the total **Atlantic swordfish** caught by longline by all countries of the world.

Year	Weight
1980	12,831
1981	10,549
1982	13,019
1983	14,023
1984	12,664
1985	14,240
1986	18,269
1987	20,022
1988	18,927
1989	15,348
1990	14,026
1991	14,208
1992	14,288
1993	15,755
1994	14,129
1995	15,615
1996	13,639
1997	12,261
1998	10,837
1999	10,754

SWORDFISH

What's the Big Deal?

Who Depends on Fish

Of the 30 countries most dependent on fish as a protein source, 26 are in the "developing world." People in developing countries such as Haiti and Lesotho depend more on fish as their main source of protein than people in industrial countries such as the U.S. and Japan. But people in industrial countries eat 40 percent of the world's fish!

How Much We Eat

Between 1995 and 1997, people in the U.S. ate an average of 46 pounds of fish per person ***(per capita)*** every year. The Japanese, biggest harvesters of fish in the world, ate a hefty 152 pounds per capita. Spain consumed 90 pounds per capita, per year, while people in Afghanistan and Ethiopia averaged only .2 pounds per person.

What Else Gets Caught

The total worldwide catch of all species of fish and shellfish is about 93 million tons. Approximately one-third of that catch is wasted—thrown back into the sea dead or dying. That's around 30 million tons—as heavy as 525 average-sized cruise ships! These unwanted marine organisms are called ***bycatch***, or incidental take.

Shrinking World Fisheries

From 1950 to 1989, there was a 300-percent increase in the amount of marine fish caught by the world fisheries. Since then, the total catch has been *decreasing* due to overfishing. Nearly every individual fishery is in decline.

FISHERY INFORMATION CARD

TUNA

What's Known about It?

How They Feed
Maybe the most amazing fact about tuna is that they're not cold-blooded, as are almost all other fish, but *warm-blooded*—which means they have a rapid metabolism and can keep their body temperatures warmer than the surrounding ocean. (Swordfish and some sharks are the same way.) This amazing adaptation allows them to swim faster and longer than cold-blooded fish, and to hunt at greater, colder depths than many other predators. This works very well in the open ocean.

The Yellowfin
Many species of tuna are caught for food, including yellowfin, bluefin, albacore, skipjack, and bonita. Here we'll concentrate on yellowfin tuna, the most significant catch in the eastern tropical Pacific. It's the species environmentalists tend to concentrate on. Remember the handout, Dolphins and Your Tuna-Fish Sandwich? Yellowfin tuna was the species protected by those "Dolphin-Safe Tuna" regulations in the 1990s.

What They Look Like
Like squids and many other marine organisms, yellowfin tuna use ***countershade coloration*** to be less visible in the water from both above and below. They have blue backs and silvery bellies; young ones may have white spots or stripes on their bellies. Yellowfin can get 6 ½ feet long.

Where They Hang Out
In the eastern tropical Pacific, yellowfin tuna move into California waters from Baja California, Mexico in late summer and fall, and return south as the northern waters cool off. Yellowfins are social fish, mostly surface-dwelling and often found away from the coast. Shoals of yellowfins like to hang around mats of drifting kelp. We're not sure why the tuna do this (because the shade attracts their prey? to hide from predators of the air?), but fishers take advantage of this behavior to lure the yellowfins under floating objects and catch them.

How Long They Live
Yellowfins live at least 5 years, but we don't know exactly how old they can really get.

Reproduction
Yellowfin tuna mature enough to ***spawn*** (lay eggs) at between 2 and 3 years old, when they're about 3 feet long. They do their spawning in warm tropical waters.

Other Tunafishy Facts
- The scientific name for yellowfin tuna is *Thunnus albacares*, from the Latin words for "tuna" and "white."
- Because tuna migrate over long distances and travel from one country's coastline to another, it's very difficult to estimate the world's tuna population.

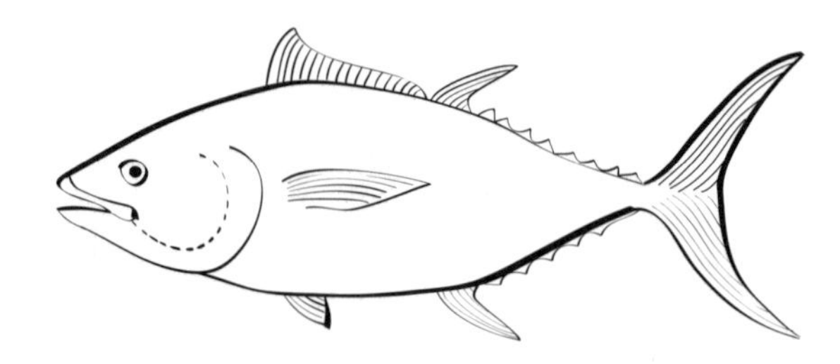

What's for Dinner?

What They Eat

Yellowfin tuna feed on fish, krill, pelagic red crab, and squids. An analysis of the stomach contents of 10 yellowfin tuna found on average that each had eaten 12 pounds of food: 7 pounds of fish, 2 pounds each of squid and krill, and one pound of pelagic red crab.

What Eats Them

As for the yellowfin's predators, well...a 157-pound yellowfin was found inside a 1,500-pound black marlin! Yellowfins are also eaten by other, larger tuna—and people in the U.S. eat more canned tuna (including yellowfin) than any other seafood.

What Tuna Canners Sell

Tuna fishers often encircle (set their nets around) dolphins to catch the yellowfin tuna that like to swim underneath. When the nets are hauled in, the dolphins are trapped inside along with the tuna, and may drown or be killed in the equipment. The death of these intelligent marine mammals is a terrible waste. Remember the "Dolphin-Safe" labels described in the handout, Dolphins and Your Tuna-Fish Sandwich? Well, the three leading U.S. tuna canners—StarKist, Bumblebee, and Chicken of the Sea, have said they'll sell only tuna caught *without* dolphin encirclement. Next time you're at a supermarket or grocery store, check a few tuna fish cans. Are they still labeled "Dolphin-Safe"?

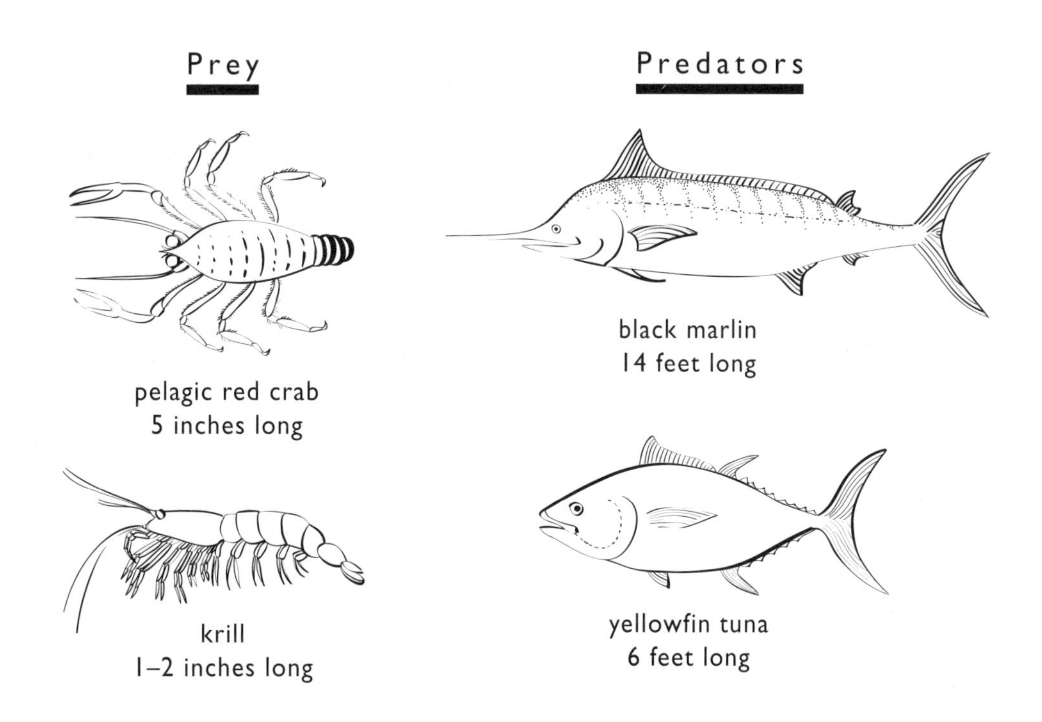

TUNA

Where in the World?

Where They're Caught

Most yellowfin tuna caught in the **Pacific** come from eastern and central tropical regions, north from the tropic of Cancer, off the coast of South America, and as far north as Southern California. In the **Atlantic**, the yellowfin tuna fishery is concentrated off the coast of Africa and along the equator. There's also a large fishery off Venezuela's western Atlantic coast.

Who Catches Them

The U.S. and Mexico catch most of the tuna landed (caught and brought to shore) in the entire **Pacific**. Others include the Philippines, China, France, Venezuela, and many South Pacific islands. In the **eastern tropical Pacific**, Mexico takes the most tuna. (See table below.) In the **central western Pacific**, the U.S., Japan, Korea, and Taiwan take the most. In the **Atlantic**, France, Ghana, and Spain do most of the fishing for yellowfin tuna.

The following table shows in metric tons (mt) the landings (size of the catch) for countries taking the most **yellowfin tuna** in the eastern tropical Pacific in 1999. (Total for 1999 was 290,000 metric tons.)

Country	Yellowfin Landings (in metric tons)
Mexico	115,000
United States of America	4,200
Ecuador	58,000
Vanuatu	19,700
Venezuela	58,000
Spain	9,000
Costa Rica	30
Colombia	12,500
Panama	6,200

The following table shows the amount of fish and shellfish **of all kinds** eaten per year, per person (per capita) in countries consuming the most and least fish per person.

Country	Fish Consumption per Capita (in pounds)
Japan	152
China	53
United States of America	46
India	10
Indonesia	40
Cambodia	20
Spain	90

How Are They Caught?

The Fishing Method

In the Pacific, yellowfin tuna are caught by a net called a ***purse seine,*** nearly a mile long. It hangs like a curtain in the water and is pulled, by a boat about 80 feet long, in a circle around the tuna. The bottom of the net is closed by pulling a "drawstring" and creating a cone, or "purse." The purse traps the fish—and anything else caught in the net (see below).

What Else Gets Caught

For reasons we don't yet fully understand, yellowfin tuna often swim under dolphins or gather under floating mats of kelp (giant seaweed) and other floating debris. Fishers take advantage of this behavior by using a ***chaser boat*** to herd dolphins into their nets to catch the tuna swimming underneath. If dolphin-escape methods aren't used (see "backing down," below) the dolphins are trapped inside when the nets are hauled in, and are injured or drown. They become wasted ***bycatch***.

Reducing Bycatch

For decades, thousands and thousands of dolphins were being trapped and killed each year along with the tuna. In 1990, though, the United States stopped buying tuna caught with those fishing methods. All tuna in U.S. stores from 1990 to 1998 were caught by *other* methods, with far less bycatch. Soon all tuna fishers in the Pacific had to change their fishing methods if they wanted to sell their catch to the U.S. They agreed to place a government observer on every tuna boat, and to use a technique called ***backing down*** to release dolphins from the nets. By the late 1990s, the number of dolphins killed had dropped by 97 percent. As you learned from the handout, Dolphins and Your Tuna-Fish Sandwich, tuna caught with dolphin-saving techniques began to be sold in cans labeled "Dolphin Safe."

A Step Backward?

Congress has been considering less-strict standards for the "Dolphin Safe" label. The government is thinking of allowing tuna cans to display the label even if the tuna *was* caught using dolphin-encircling methods—as long as shipboard observers don't actually see dolphins killed or injured. Supporters of this change believe it would allow more manufacturers to use the "Dolphin Safe" label while still making sure dolphins aren't harmed. Opponents argue that dolphins will still die or be harmed as they're chased over miles of open ocean and encircled in nets up to a mile long. In 2000, a federal judge ruled that more studies and review of the data need to be done. What do you think?

Other Measures

In 2000, the U.S. also decided not to import yellowfin tuna from the eastern tropical Pacific—unless the nations fishing there can prove the tuna was caught using dolphin-safe measures. Those nations include many Central American countries (most importantly Mexico, but also Belize, Guatemala, and El Salvador), Venezuela, Colombia, and Spain.

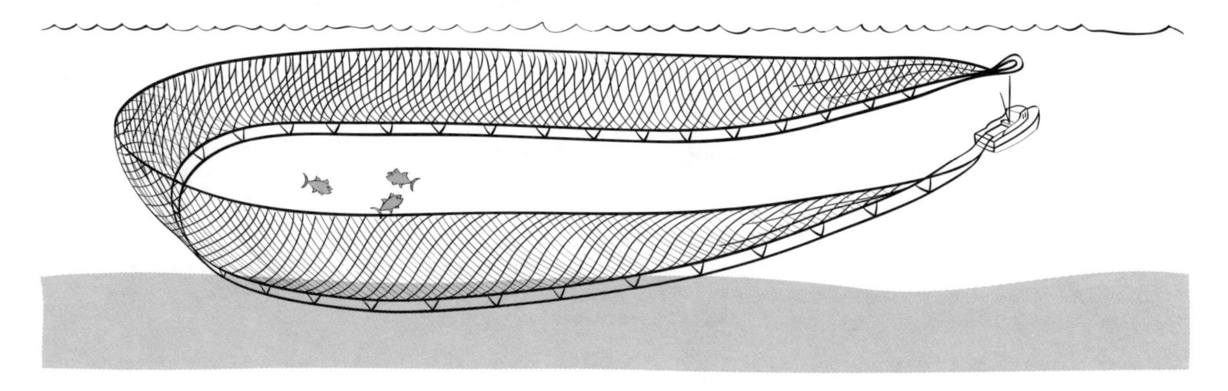

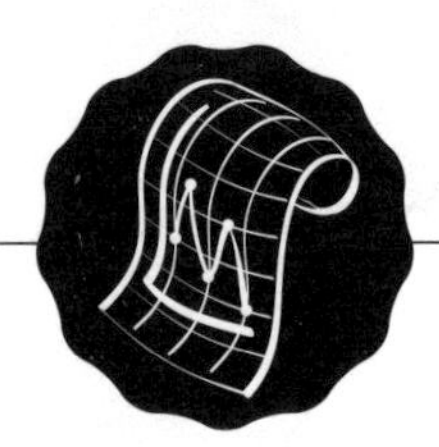

TUNA

What's Happening with the Fishery?

The Kind of Tuna We Buy

The U.S. eats the most canned tuna in the world. Many species of tuna are caught, but the most economically important are yellowfin, bluefin, albacore, skipjack, and bonita. All these kinds of tuna end up in cans except the very expensive bluefin, which is only sold whole or in pieces. Albacore is the highest quality of canned tuna, and is labeled "white meat tuna"; yellowfin is canned as "chunk light tuna."

Note: Bluefin tuna can't be sold in the U.S. because the fish has such high mercury content that it's considered unsafe to eat by U.S. standards. (It would probably be too expensive for the U.S. market anyway!)

Setting Limits

The amount of fishing that can occur while still leaving enough animals to reproduce is called the ***maximum sustainable yield (MSY).*** The MSY for yellowfin tuna in the eastern tropical Pacific is about 280,000 metric tons. But in 1999 the average catch of yellowfin for the eastern tropical Pacific was 296,392 metric tons. It appears the yellowfin is being fished at or above the level it can sustain. (In the Atlantic, yellowfins may be on the brink of being overfished. The Atlantic yellowfin declined by 30 percent between 1988 and 1998.)

Managing the Fishery

Most worldwide tuna populations have declined due to overfishing. Even if the number of boats fishing for tuna were kept at a constant level, the fishing *capabilities* of the fleet are increasing as fishing gear and techniques become more and more efficient. No overall management plan covers the tuna fishery in international waters—or even in all U.S. waters. Instead, a number of separate organizations and treaties, with limited power over limited areas, try to manage the fishery. Two new international agreements (the United Nations Highly Migratory Fish Stocks and the Code of Conduct for Responsible Fishing) may help protect the tuna.

The table below shows in metric tons (mt) the total eastern tropical Pacific **yellowfin tuna** caught by all countries of the world.

Year	Weight	Year	Weight
1980	175,000	1990	348,000
1981	170,000	1991	300,000
1982	150,000	1992	325,000
1983	125,000	1993	271,000
1984	150,000	1994	280,000
1985	200,000	1995	225,000
1986	285,000	1996	240,000
1987	286,000	1997	250,00
1988	320,000	1998	270,000
1989	325,000	1999	290,000

Footnote: In the 1980s, tuna fishing in the eastern tropical Pacific became less profitable. Many U.S. fishers moved away or quit, leaving Mexico the dominant fleet in these waters. In the '90s, when dolphin protection laws put added pressure on U.S. purse seiners, the U.S. fleet virtually disappeared.

What's the Big Deal?

Who Depends on Fish

Of the 30 countries most dependent on fish as a protein source, 26 are in the "developing world." People in developing countries such as Haiti and Lesotho depend more on fish as their main source of protein than people in industrial countries such as the U.S. and Japan. But people in industrial countries eat 40 percent of the world's fish!

How Much We Eat

Between 1995 and 1997, people in the U.S. ate an average of 46 pounds of fish per person ***(per capita)*** per year. The Japanese, biggest harvesters of fish in the world, ate a hefty 152 pounds per capita. Spain consumed 90 pounds per capita, per year, while people in Afghanistan and Ethiopia averaged only .2 pounds per person.

What Else Gets Caught

The total worldwide catch of all species of fish and shellfish is about 93 million tons. Approximately one-third of that catch is wasted—thrown back into the sea dead or dying. That's around 30 million tons—as heavy as 525 average-sized cruise ships! These unwanted marine organisms are called ***bycatch***, or incidental take.

Shrinking World Fisheries

From 1950 to 1989, there was a 300-percent increase in the amount of marine fish caught by the world fisheries. Since then, the total catch has been *decreasing* due to overfishing. Nearly every individual fishery is in decline.

SQUID

What's Known about It?

Mystery Mollusk

Squids are open-ocean mollusks related to octopuses, snails, and clams. Every spring and summer, California market squids (the most popular commercial squids) come together in large numbers, or ***shoals***, to breed. We still know very little about the biology or population size (large, we think) of the market squid. We don't know what they do in the open ocean, or where they go before they come together in mating shoals. Researchers are now trying to study them through their entire life cycle (rather than just when adults appear in large groups to breed), to better manage the fishery.

Reproduction

Female squids ***spawn*** (lay eggs) primarily at night. They produce strands of "egg capsules" covered in a jelly-like material, and attach them in clusters to the sandy or muddy bottom. Each capsule contains as many as 300 eggs. About eight weeks after hatching, the young squids start to swim in shoals.

How Long They Live

The males and females die soon after spawning, when they're less than one year old.

What They Look Like

Like many other marine organisms, squids use ***countershade coloration*** to be less visible in the water from both above and below, generally appearing paler from below and darker from above. They can change color at will—and very rapidly—in reaction to sudden events or to camouflage themselves. They have eight arms and two, longer tentacles, and their brains and eyes are very large.

Other Squiddy Facts

- The scientific name for California market squid is *Loligo opalescens*. The first word probably comes from the name of a squid relative, the cuttlefish; *opalescens* is "iridescent" (like a polished opal).
- More than 500 species of squid inhabit the world's ocean—but fewer than 12 species make up 90 percent of the global catch!

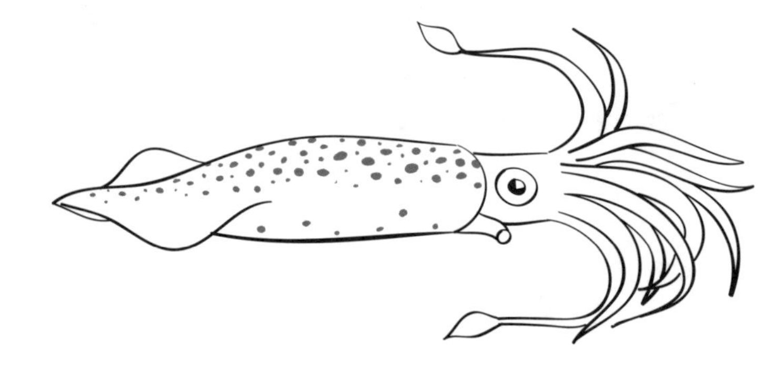

SQUID

What's for Dinner?

What They Eat

Squids capture and eat many kinds of prey, including worms, crabs, shrimp, fish, and even other squids. Of 100 market squids sampled for their stomach contents, 10 had eaten small fish (mainly anchovies), 50 had eaten shrimp and krill, 10 had eaten pelagic red crabs, and 30 had eaten other squids.

What Eats Them

Dolphins, sperm whales, pilot whales, porpoises, seals, sea lions, sea otters, salmon, swordfish, tuna, sharks, and seabirds such as shearwaters and gulls depend on squids for food. Squid is also an important "people food" in many parts of the world. New, larger boats can deliver fresher squid to market, and more North Americans are eating squid than ever before. In North American restaurant menus and stores, squid is often called *calamari*—which is simply the Italian word for squid!

Squid Predators at Risk

Scientists, environmentalists, and sport fishers are very concerned about the survival of all the marine mammals, fish, and birds that depend on squids for food. Any decline in the squid fishery affects the survival of these species too. A sustainable fishery is as important to the ocean ecosystem as it is to the fishing industry.

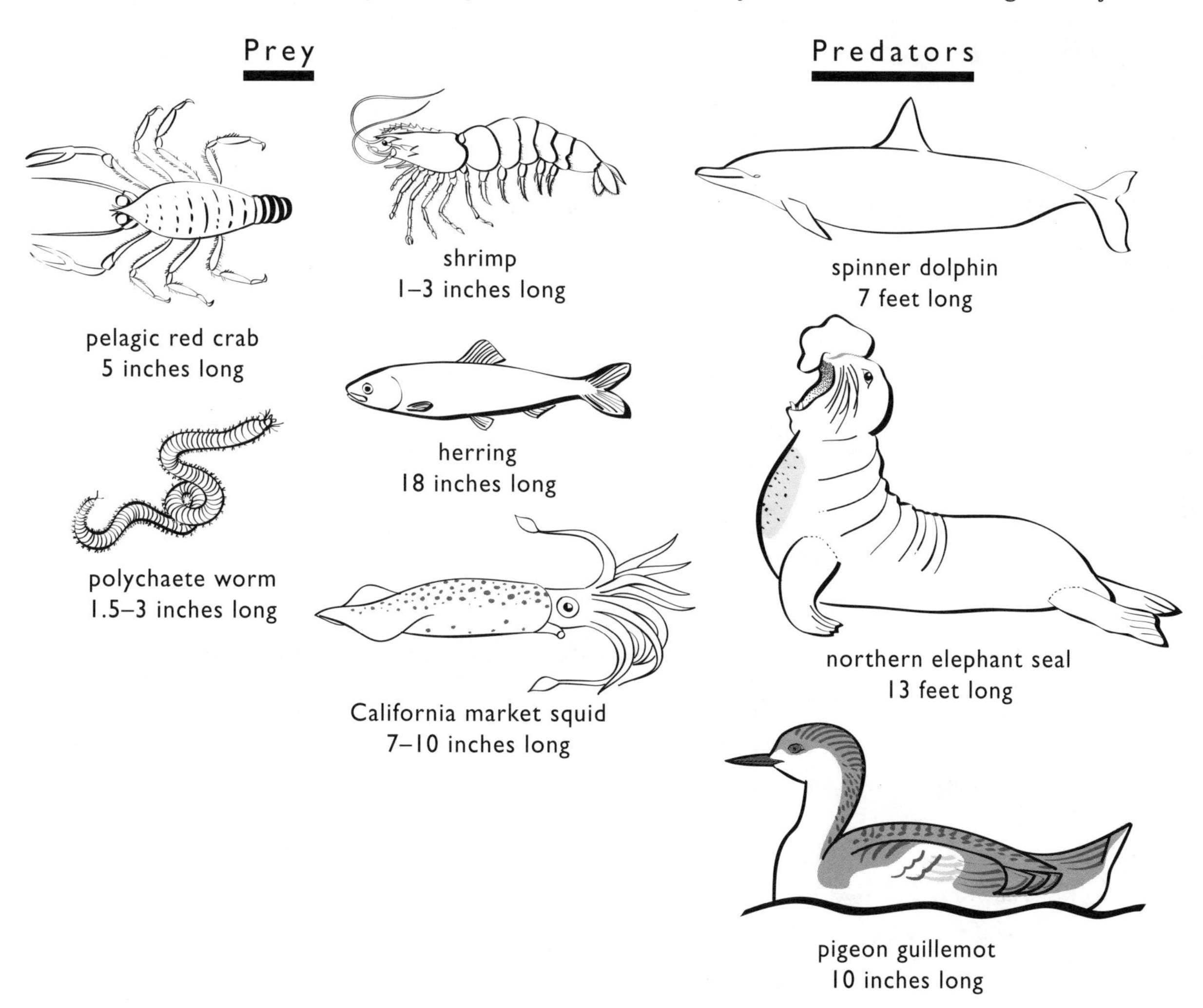

FISHERY INFORMATION CARD

SQUID

Where in the World?

Where They're Caught

Squid is fished throughout the ocean basins of the world. Squids are caught off the tip of India, around New Zealand, along the entire western Pacific from the Gulf of Siam west to the Philippines, and north throughout the Sea of Japan to Vladivostok, Russia. In the U.S., common market squids are caught from southeast Alaska to Baja California, Mexico. Most are caught in Central and Southern California, especially around the Channel Islands. In Atlantic coastal waters, squids are caught from the Gulf of St. Lawrence and Newfoundland in eastern Canada, south to Cape Hatteras in North Carolina. Squids are also caught around the Falkland Islands, off the coasts of Argentina, Peru, and Ecuador in South America, and off the southeastern coast of South Africa.

Who Catches Them

Japan catches 80 percent of the world's squid, and eats the most squid of any country in the world. Korea and Argentina also catch huge amounts of squid. In the U.S., the squid catch has increased dramatically since 1992, and California now catches 67 percent of the U.S. total. The table below shows the catch of cephalopods—mostly squids—by several countries.

The following table shows in metric tons (mt) the landings (size of the catch) for countries taking the most **cephalopods (especially squids)** of all kinds in 1999. (Total for 1999 was 2,314,000 metric tons.)

Country	Cephalopod Landings (in metric tons)
Japan	650,000
United States of America	114,000
Korea	450,000
Argentina	400,000
Taiwan	200,000
Others (New Zealand, Philippines, India, South Africa, Russia....)	500,000

The following table shows the amount of fish and shellfish **of all kinds** eaten per year, per person (per capita) in countries consuming the most and least fish per person.

Country	Fish Consumption per Capita (in pounds)
Japan	152
China	53
United States of America	46
India	10
Indonesia	40
Cambodia	20
Spain	90

SQUID

How Are They Caught?

The Fishing Methods

1. In the U.S., 90 percent of squid fishers use a kind of net called a ***purse seine,*** nearly a mile long, to make their catch. When squids come together in big groups to ***spawn*** (lay eggs) on spring and summer nights, fishers shine large, bright lights over the water to attract the squids. The purse seine, which hangs like a curtain through the water, is then pulled by a boat about 80 feet long in a circle around the squids. Fishers close the bottom of the net by pulling a "drawstring" and creating a cone, or "purse." The purse traps the squids (and any unwanted organisms, or ***bycatch***) in the net. The entire net is then hauled in.

2. Japan catches more squids than any country in the world, and has the most high-tech squid-fishing methods. The Japanese use a technique called ***jigging***. It uses many lures, each with two or three rows of hooks in a ring fastened to the fishing line. Either by hand or by machine, the lures are "jigged" (wiggled) in the water from a boat about 100 feet long to attract the squids. Jigging catches the common Japanese squid, but can also catch much larger and more powerful species of squid.

One Example of Managing the Fishery

The Falkland Islands use money from fishing boat license fees to conserve and manage their squid fishery by:

- reducing the number of boats allowed to fish for squids;
- decreasing the number of squids caught as ***bycatch*** in other fisheries; and
- requiring boats to report their catches and carry observers on board.

These efforts have paid off; the Falkland Islands appear to have achieved a sustained fishery—meaning the squids can continue to reproduce despite fishing pressure. The Falklands have also created a Marine Protected Area where most of the squid's annual life cycle takes place. This allows the squids to develop and mature undisturbed.

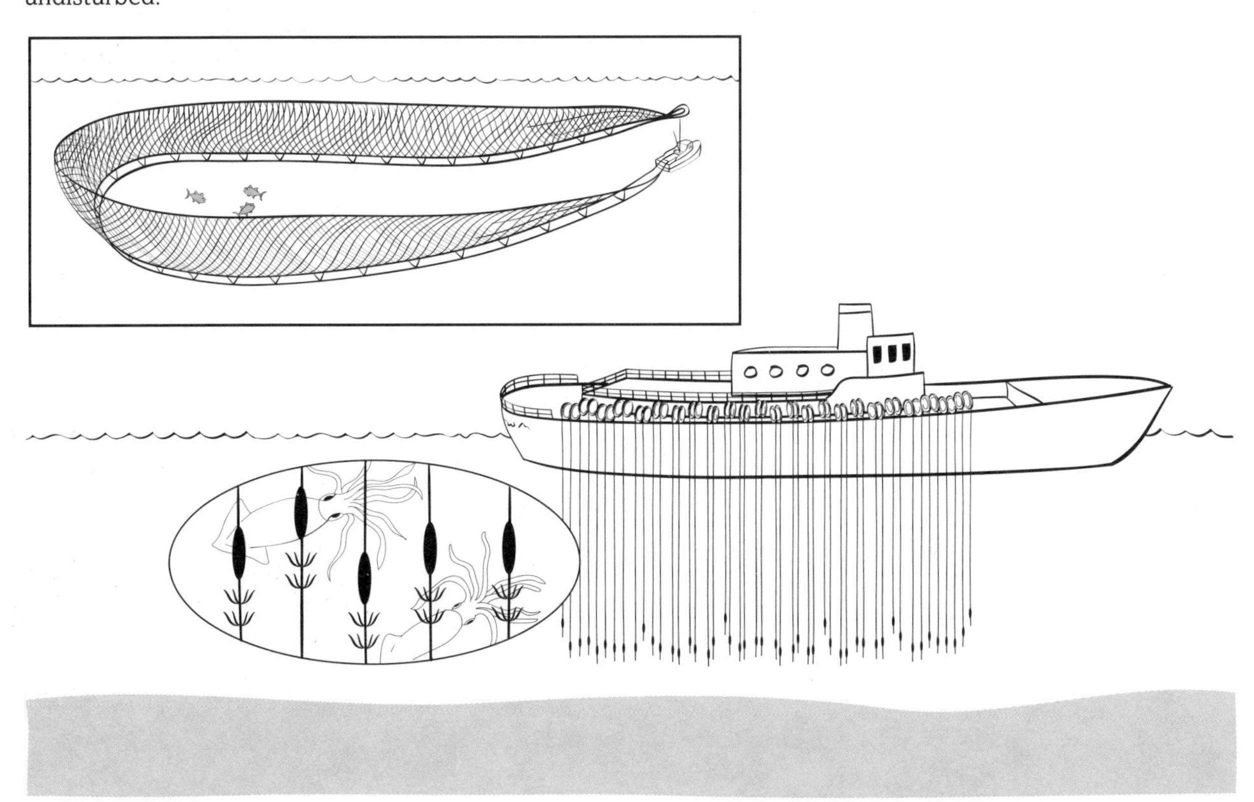

SQUID

What's Happening with the Fishery?

Squids on the Edge of Disaster

From 1997 to 1998, market-squid landings (the catch that's brought to shore) decreased drastically, dropping by 67,000 tons in a single season. The combination of a number of factors threatened disaster for market squids: insufficient scientific information, an unregulated fishery, and unfavorable environmental conditions (this was an *El Niño* year, when weather disruption affected oceanic productivity). We just don't know how much more fishing pressure squids can handle. The number of squid boats is growing, and powerful new fishing techniques are spreading throughout the fishery.

Setting Limits

In 1999 the U.S. government enacted the Final Rule for Coastal Oceanic Species, making the market squid a "monitored" species. This allowed fishery managers to monitor its population by studying commercial catch data. The idea was to establish a ***maximum sustainable yield (MSY)*** (the number of squids that can be caught while still leaving enough animals to reproduce), which was unknown for squids at the time. Many California fishers wanted to stop any new boats from fishing for squids, because they were afraid there wouldn't be enough squids for everyone. Fishers also wanted new laws to protect the fishery—they knew that if the squid fishery collapsed, there were no other fisheries left in California to go into.

Managing the Fishery

As squids were monitored, the relationship of squid-fishing methods to size, sex, and age of the catch became clearer. Based on this information, the California Department of Fish and Game was able to start a real squid management plan in May 2001, to establish and ensure sustainable yields. New regulations:

- limit how many lights can be used on boats to attract squids;
- limit how strong the lights can be; and
- require lamp shields to direct the light down into the water so birds won't be attracted to the boats and get caught in the nets.

The management plan also includes earlier regulations that:

- close the squid fishery (make it off-limits) on weekends, so squids can ***spawn*** (lay eggs) undisturbed for several days every month;
- require market-squid fishers to keep logbooks on their catches (providing critical information on fishing practices).

More Management Needed

We still need to research the best plan to manage market squids around the planet. Squids in South African waters and around the Falkland Islands are now carefully managed. In those areas, squid biology is studied, fishing is restricted to certain places at certain times of year, and the number of boats is strictly limited. In this species, which may live less than a year, keeping enough squids for the fishery will depend almost entirely on whether their young get the chance to survive.

The table below shows in metric tons (mt) the total **California market squid** caught in the U.S. between 1981 and 1999.

Year	Weight	Year	Weight	Year	Weight
1981	23,510	1988	37,232	1994	70,252
1982	16,308	1989	40,893	1995	80,561
1983	1,824	1990	37,389	1996	70,329
1984	564	1991	13,110	1997	2,894
1985	10,276	1992	42,830	1998	91,519
1986	21,278	1993	55,383	1999	115,000
1987	19,984				

What's the Big Deal?

Who Depends on Fish

Of the 30 countries most dependent on fish as a protein source, 26 are in the "developing world." People in developing countries such as Haiti and Lesotho depend more on fish as their main source of protein than people in industrial countries such as the U.S. and Japan. But people in industrial countries eat 40 percent of the world's fish!

How Much We Eat

Between 1995 and 1997, people in the U.S. ate an average of 46 pounds of fish per person ***(per capita)*** per year. The Japanese, biggest harvesters of fish in the world, ate a hefty 152 pounds per capita. Spain consumed 90 pounds per capita, per year, while people in Afghanistan and Ethiopia averaged only .2 pounds per person.

What Else Gets Caught

The total worldwide catch of all fishery species is about 93 million tons. Approximately one-third of that catch is wasted—thrown back into the sea dead or dying. That's around 30 million tons—as heavy as 525 average-sized cruise ships! These unwanted marine organisms are called ***bycatch***, or incidental take.

Shrinking World Fisheries

From 1950 to 1989, there was a 300-percent increase in the amount of marine fish caught by the world fisheries. Since then, the total catch has been *decreasing* due to overfishing. Nearly every individual fishery is in decline.

FISHERY INFORMATION CARD

WALLEYE POLLOCK

What's Known about It?

World's Most Popular Fish
Walleye pollock is the most-fished species in the world. (Few people know this fish by name, but if you eat fish sticks, you're probably eating pollock!) A vast number of ships from many nations fish for walleye pollock in both the Atlantic and Pacific ocean basins.

What They Look Like
Walleye pollock are olive-green or brown on the back and silvery on the sides. They have three dorsal fins and two anal fins, and get about 3 feet long. Young ones have two to three yellow stripes on the sides.

Where They Hang Out
Walleye pollock swim in shoals and spend most of their time along the outer continental shelf. They live in mid-water and on the ocean floor, as deep as 3,000 feet below the surface. (Fish that live on the bottom are called "groundfish.")

How Long They Live
This species lives at least 14 years.

Reproduction
Walleye pollock migrate from the continental shelf to shallow water during the spring and summer to ***spawn*** (lay eggs) and feed, and return to deeper water in the fall and winter. They spawn at 3 years of age and lay from 20,000 to 100,000 eggs at a time. Spawning season varies according to where they live. In the Bering Sea, they spawn from February to June (May is the peak); in the Gulf of Alaska most spawning occurs in March and April.

Another Pollocky Fact
- The scientific name for walleye pollock is *Theragra chalcogramma*, meaning "food beast." It's thought they were called that because so many marine mammals depend on them for food.

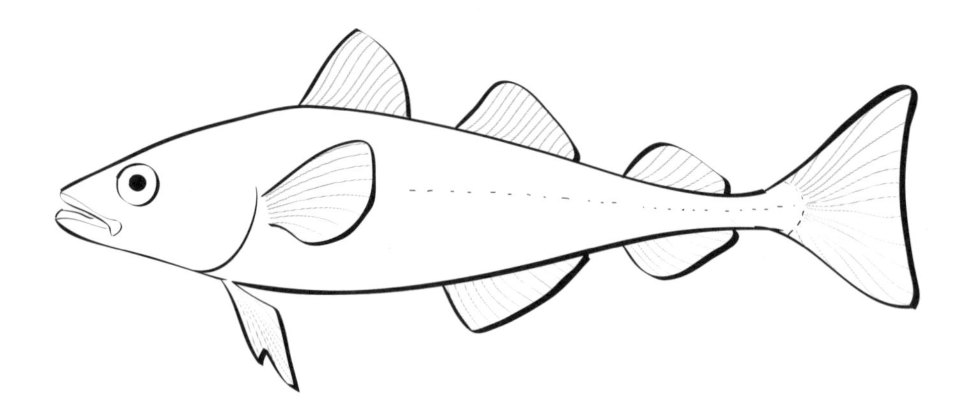

WALLEYE POLLOCK

What's for Dinner?

What They Eat

These fish are big eaters. Walleye pollock are "groundfish," or bottom-living predators, and their main foods are large zooplankton (crustaceans, etc.), and small zooplankton (copepods, etc.). They also eat small fish—including juvenile pollock!—as well as small amounts of worms and clams. Analysis of 100 pollock showed their combined stomach contents to be 10 pounds of food: 4.5 pounds of crustaceans; 3.5 pounds of copepods; 1.5 pounds of juvenile pollock; and .5 pound of worms and clams.

What Eats Them

Walleye pollock are an extremely important prey item for seabirds and fish (including salmon), and for many marine mammals such as Steller sea lions and harbor seals. Marine mammals consume about the same amount of walleye pollock each year as humans catch commercially. People also eat a lot of pollock, mostly in the form of roe (the eggs), fish sticks, or *surimi* (fake crab, lobster, and scallops). *Surimi* makes up about one-fourth of all the seafood eaten in Japan.

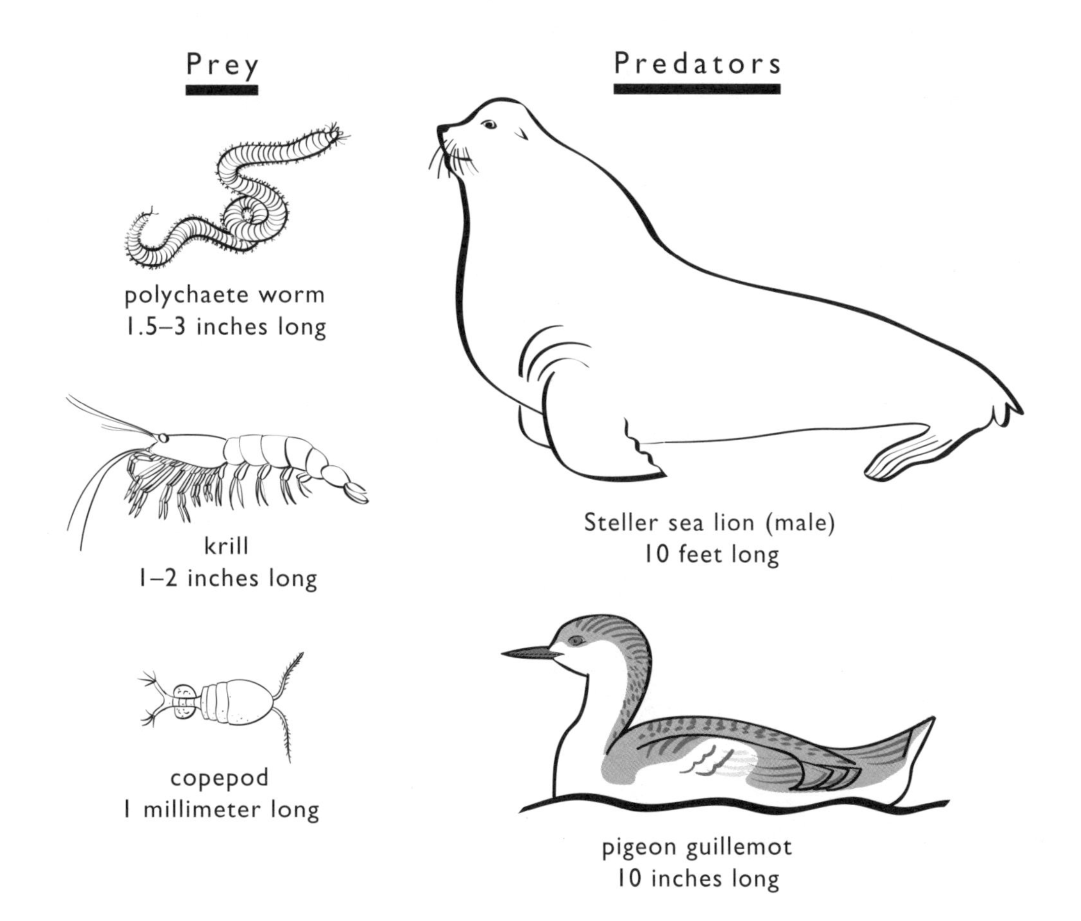

WALLEYE POLLOCK

Where in the World?

Where They're Caught

The walleye pollock fishery covers much of the North Pacific. It's concentrated in the Gulf of Alaska and in the Bering Sea, but stretches from the coast of British Columbia in western Canada all the way to the Asian mainland. Walleye pollock can be found from the Chukchi Sea (north of the Bering Sea) to the southern Sea of Japan, then east to the Aleutian Islands and along the continental shelf, throughout the Gulf of Alaska as far south as Central California.

Who Catches Them

This is the largest fishery in the world. Russia catches the most walleye pollock (by weight), with Japan, Korea, China, and the U.S. scooping up most of the rest. This pollock is the most-harvested fish in the U.S.; Alaska's combined fisheries are much larger than those of other states—and walleye pollock is one of the major reasons.

The following table shows in metric tons (mt) the landings (size of the catch) in 1998 for the countries taking the most **walleye pollock** in the North Pacific. (Total for 1998 was 4,687,587 metric tons.)

Country	Pollock Landings (in metric tons)
Russia	2,076,537
Japan	373,230
Korea	300,500
China	141,442
United States of America	1,711,962
Poland	81,889

The following table shows the amount of fish and shellfish **of all kinds** eaten per year, per person (per capita) in countries consuming the most and least fish per person.

Country	Fish Consumption per Capita (in pounds)
Japan	152
China	53
United States of America	46
India	10
Indonesia	40
Cambodia	20
Spain	90

WALLEYE POLLOCK

How Are They Caught?

The Fishing Method

Walleye pollock are caught with ***factory trawlers***, massive ships up to 550 feet long—the largest fishing vessels in the world. They're really "factories at sea," catching the pollock with huge nets and processing and storing the fish on board. The largest of them drag ***trawls*** (nets) as long as four 360-foot football fields—each big enough to hold several jumbo jets and capable of capturing over 400 tons of fish in a single haul. Factory trawlers can stay out at sea for months at a time without returning to port, as they trawl and process fish on board around the clock, seven days a week.

What Else Gets Caught

The gigantic nets of factory trawlers drag along the ocean floor and scoop up everything in their path. Bottom trawling is the most efficient way to catch groundfish like pollock—but it also catches tons of other, non-target marine life. Just about everything the fishers bring up is dead; the unwanted organisms, called ***bycatch***, are thrown back into the sea as waste. When lots of bycatch species such as crab and halibut are caught accidentally in the walleye pollock fishery, the crab and halibut fisheries get smaller and smaller. Trawling for walleye pollock may prevent those other fisheries (already diminishing) from making a comeback.

Reducing Bycatch

Several regulations address the problem of bycatch from trawling. Here are two of those rules:

- Valuable bycatch such as halibut cannot be kept and sold. This ensures that fishers must fish more carefully for their target species, and aren't allowed to benefit from their bycatch.
- Observers must be placed on all fishing vessels to monitor bycatch and make sure the boats only take as much total catch as the law allows.

Damage to the Ocean Floor

When factory trawlers drag their huge weighted nets over the ocean floor, they seriously damage the sea bed. This destruction of the ocean bottom alters the habitat that walleye pollock—and many other organisms—need to survive.

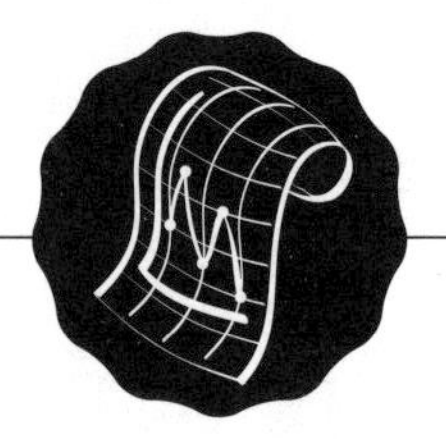

WALLEYE POLLOCK

What's Happening with the Fishery?

Too Many, Too Big

Huge numbers of ships from around the world fish for walleye pollock in the Atlantic and Pacific ocean basins. These gigantic factory trawlers can stay at sea for long periods of time, and they catch enormous quantities of pollock and unwanted bycatch. In the U.S., the powerful North Pacific pollock trawling fleet is so large that it actually could catch two to three times more pollock than it's allowed to catch each year! The ever-increasing trawler fleet is rapidly overfishing even the North Pacific's productive seas.

Setting Limits—and Helping the Pollock

The amount of fishing that can occur while still leaving enough animals to reproduce is called the ***maximum sustainable yield*** **(MSY).** For walleye pollock caught by the U.S., the MSY is about 1,880,000 metric tons—yet pollock fishers stop fishing before reaching the pollock's maximum sustainable yield. That's because in the walleye pollock fishery, the government counts bycatch numbers as part of the maximum sustainable yield. The more bycatch the fishers get, the sooner they have to stop fishing for what they really want—walleye pollock. This is one way to protect pollock from being overfished.

Biology at Work?

In several regions where walleye pollock were once plentiful and heavily fished, their numbers are now extremely low. Some experts believe the decline in pollock in these areas isn't because of overfishing but because of a natural phenomenon; many walleye pollock are getting old at the same time, and that group is dying off. The experts hope the arrival of a new generation of pollock will replenish the population. If that doesn't happen, though, fishery managers will be forced to lower the number of walleye pollock the fishers can catch, to protect the remaining population.

Other Animals Are Affected Too

Overharvesting of walleye pollock has also affected the marine mammals, fish, and seabirds that depend on them for food. Steller sea lions, an endangered species that feeds on walleye pollock, have decreased alarmingly since the 1970s. Many experts think it's largely because of competition for pollock between sea lions and human fishers. Other marine mammals, such as fur seals and harbor seals, are severely threatened too; and many seabirds such as murres and kitiwakes are declining throughout the Gulf of Alaska. New regulations may help reduce fishery competition with these endangered animals. The new rules make important sea lion breeding and feeding habitats off-limits for fishing, and reduce the number of walleye pollock that can be caught in other habitats.

Other Management Measures

Researchers and walleye pollock fishers have pushed to design regulations and better gear to reduce the bycatch of such non-targeted species as halibut, crab, salmon, and herring. One new regulation guarantees that every boat will get a fair share of the total walleye pollock catch for the year. This encourages fishers to take more time to fish for the right size of pollock, rather than rushing to snatch lots of fish of any size.

The table below shows in metric tons (mt) the total **walleye pollock** caught (90% by trawlers) in Alaska by the U.S.

Year	Weight	Year	Weight	Year	Weight
1980	15,770	1987	1,326,000	1994	1,496,000
1981	138,403	1988	1,405,000	1995	1,402,000
1982	259,892	1989	1,340,800	1996	1,274,000
1983	430,641	1990	1,431,000	1997	1,241,000
1984	1,485,000	1991	1,737,000	1998	1,250,000
1985	1,513,000	1992	1,534,000	1999	1,086,000
1986	1,330,000	1993	1,494,000		

WALLEYE POLLOCK

What's the Big Deal?

Who Depends on Fish

Of the 30 countries most dependent on fish as a protein source, 26 are in the "developing world." People in developing countries such as Haiti and Lesotho depend more on fish as their main source of protein than people in industrial countries such as the U.S. and Japan. But people in industrial countries eat 40 percent of the world's fish!

How Much We Eat

Between 1995 and 1997, people in the U.S. ate an average of 46 pounds of fish per person ***(per capita)*** every year. The Japanese, biggest harvesters of fish in the world, ate a hefty 152 pounds per capita. Spain consumed 90 pounds per capita, per year, while people in Afghanistan and Ethiopia averaged only .2 pounds per person.

What Else Gets Caught

The total worldwide catch of all species of fish and shellfish is about 93 million tons. Approximately one-third of that catch is wasted—thrown back into the sea dead or dying. That's around 30 million tons—as heavy as 525 average-sized cruise ships! These unwanted marine organisms are called ***bycatch***, or incidental take.

Shrinking World Fisheries

From 1950 to 1989, there was a 300-percent increase in the amount of marine fish caught by the world fisheries. Since then, the total catch has been *decreasing* due to overfishing. Nearly every individual fishery is in decline.

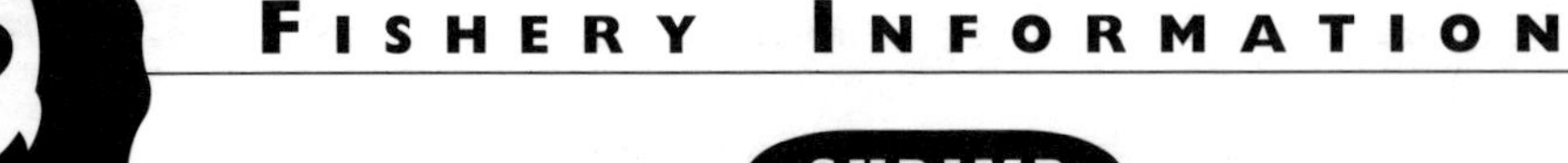

FISHERY INFORMATION CARD

SHRIMP

What's Known about It?

What Kind We Buy

Three species of shrimp are caught commercially in the southeastern U.S. and the Gulf of Mexico—brown, white, and pink. Brown shrimp are the most valuable, and the species we'll concentrate on here.

Reproduction and Life Cycle

Shrimp ***spawn*** (lay eggs) offshore in the Gulf of Mexico or southeastern Atlantic. One female shrimp can release 100,000 to 1,000,000 eggs at a time, and these hatch within 24 hours. The tiny offspring are quickly carried toward shore by tides and currents to estuaries (where rivers meet the ocean) and other wetlands along the coast, which offer lots of food and some protection from predators. There, the young shrimp grow and begin living on the bottom. If conditions are good, they grow rapidly and soon move to the deeper water of the bays. From there, the shrimp usually migrate back to the Gulf of Mexico and other southeastern waters, where they mature and complete their life cycles. Most shrimp will spend the rest of their lives in the Gulf.

How Long They Live

Shrimp mature during their first year, and—if not eaten by fish or caught by fishers—can live to be 2 years old.

What They look Like

Brown shrimp have medium-length antennae and grooves down both sides of the spine on the head and tail. They can grow over 1 ½ inches a month, and reach 9 inches long.

Other Shrimpy Facts

- The scientific name for brown shrimp is *Penaeus aztecus.* One theory is that they're named for the Penaeus River in northwest Greece and for the Aztec people of Mexico; perhaps both cultures harvested this kind of shrimp at one time–or still do!
- In order to grow, shrimp have to cast off their shell (exoskeleton) and get larger before their new shell hardens. This process is called *molting.*

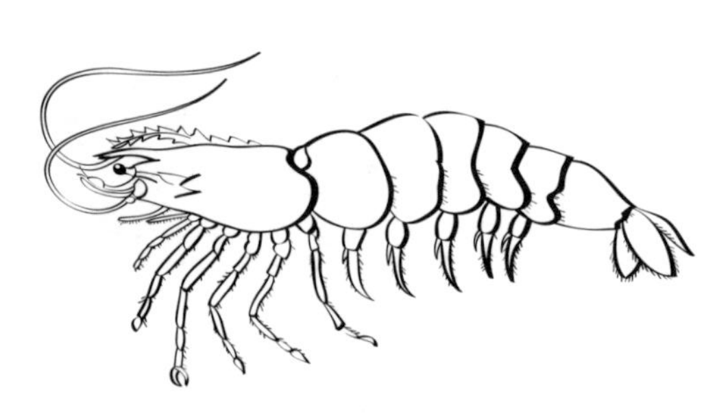

What's for Dinner?

What They Eat

Shrimp eat ***detritus***—rich organic nutrients from the breakdown of algae, animals, and microorganisms. Juvenile (young) shrimp find this detritus in estuaries, where rivers meet the ocean along the coast. As the shrimp develop and migrate to the open ocean, they eat detritus from organisms that live out at sea. Shrimp also eat a variety of small worms called polychaetes and nematodes.

What Eats Them

More and more people all over the world eat shrimp. Japan and the U.S. prefer tropical (warm-water) shrimp, while Europeans prefer cold-water species. Most (80 percent) of the shrimp in the U.S. is bought by restaurants. Many important game fish like to eat shrimp, and the sale of live shrimp to be used as bait by sport fishers is a big business. Spotted sea trout, red drum, red snapper, and many other fish depend on shrimp as their main prey. One 25-pound red snapper had 3 pounds of food in its stomach: .5 pounds of fish, .75 pounds of squid, 1.5 pounds of shrimp, and .25 pounds of other crustaceans.

"Wild-Caught" versus "Farmed"

The shrimp the world buys today may be either "wild-caught" (fished in the open ocean) or "farmed"—a method called ***aquaculture***. Because coastal marshes, estuaries, and shallow bays are the best habitat for developing shrimp, the Gulf of Mexico has the most productive *wild-caught* shrimp fishery in the U.S. Most *farmed* shrimp come from Asia and Latin America.

Why We Farm Shrimp

As world demand for shrimp increases, the fishery for wild-caught shrimp is shrinking. As a result of this overharvest, we've turned to factory-style coastal shrimp farms—aquaculture—to provide over one-fourth of the shrimp the world now eats.

The Problem with Aquaculture

When the ocean floor is raked clean to make room for the artificial farms used in aquaculture, great damage is done to natural habitats. In addition, aquaculture depends on other fish products to feed its shrimp—and that reduces the species in other fisheries.

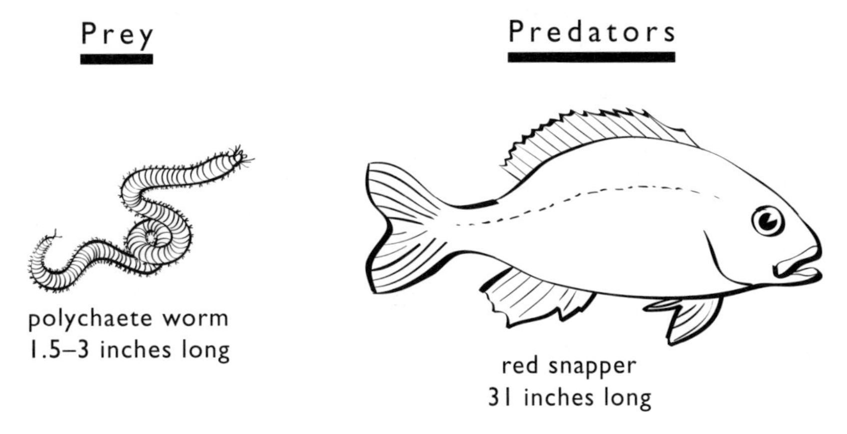

SHRIMP

Where in the World?

Where They're Caught

Three-fourths of the shrimp caught by the U.S. come from the Gulf of Mexico, one of our nation's richest fishing areas. Brown shrimp, the most valuable, are caught mainly off the Texas-Louisiana coast. Other wild-caught shrimp eaten in the U.S. come from neighboring Latin American or Caribbean countries. Shrimp are also harvested off the entire U.S. Pacific coast, where the fishery is considered to be fully fished (fished to its very limits). Even with all this fishing, however, the amount of wild-caught shrimp isn't enough to satisfy the growing world taste for this delicacy. To make up for this so-called "shortfall" of shrimp, between one-half and two-thirds of the shrimp we eat in the U.S. now come from shrimp "farms" located in Asia and Latin America. Shrimp are also caught or farmed in the coastal waters of Indonesia, Thailand, Australia, and the Philippines.

Who Catches Them

Most of the wild-caught shrimp in the world are caught by countries with fishing fleets operating in their own near-shore waters. These include China, India, the United States, Japan, Mexico, and the Philippines, and several Central and South American countries. Western European countries such as Denmark, Norway, and Iceland catch the cold-water variety of shrimp off their coastlines. Japan and the United States catch a lot of shrimp from their own portions of the continental shelf, but shrimp are in such high demand in these countries that they buy at least as much shrimp from other countries as they capture themselves. The U.S. eats more shrimp (by weight) than any other country in the world. (The amount of shrimp eaten in the U.S. has doubled in the last 10 years to one billion pounds per year!) Japan as a whole buys the second-greatest amount of shrimp (by weight) in the world, but the Japanese people actually *eat* the most shrimp *per person* (per capita) in the world.

The following table shows in metric tons (mt) the landings (size of the catch) in 1998 for the countries taking the most **shrimp**. (Total for 1998 was 3,764,934 metric tons.)

Country	Shrimp Landings (in metric tons)	Country	Shrimp Landings (in metric tons)
China	991,871	Iceland	62,727
Central/South America	448,600	Oceania	31,977
India	385,403	Japan	29,345
U.S.	127,991	Spain	27,818
Mexico	90,335	Netherlands and Denmark	22,602
Philippines	71,608		

The following table shows the amount of fish and shellfish **of all kinds** eaten per year, per person (per capita) in countries consuming the most and least fish per person.

Country	Fish Consumption per Capita (in pounds)
Japan	152
China	53
USA	46
India	10
Indonesia	40
Cambodia	20
Spain	90

How Are They Caught?

The Fishing Method

Shrimp are usually caught with ***trawls***. A trawl is a funnel-shaped net pulled through the water by a ***trawler,* or** fishing boat, that's upwards of 65 feet long. The weighted net forms a cone that tapers to a narrow end called the tail, or "cod-end," where the catch accumulates. At the front (open) end, the trawl's mouth is held open by wooden doors, and lines run from the shrimp boat to each door. As the shrimp boat tows the trawl over the sea bottom where the shrimp are, the trawl is held open by the doors. Shrimp and bottom fish are scooped into the open trawl (about 50 feet wide at the mouth) and pushed up into the cod-end. When the net is brought back aboard the boat with the help of winches (machines that pull), a line that holds the cod-end closed is released and the catch falls to the deck.

What Else Gets Caught

Bottom trawling may be the most efficient way to catch shrimp, but it catches and kills everything else in its path, too—including sea turtles and fish that live on the bottom. When the nets are hauled in and opened on the deck, not only shrimp but many other marine animals fall out. The fishers go through a long and difficult process to separate the shrimp from all the other, unwanted organisms, or ***bycatch***. The only way to reduce this bycatch is to fit the nets with devices that exclude some of the unwanted catch.

World's Most Wasteful Fishery

Shrimp trawling wastes more marine organisms than any other method of fishing on Earth. The shrimp fishery produces less than two percent of all the world's seafood, but is responsible for 35 percent of the bycatch of all the world's fisheries. For every pound of shrimp caught and kept, up to 10 pounds of other marine organisms are caught and thrown away. That means that 84 percent (by weight) of everything caught in the shrimp fishery is bycatch, and goes to waste.

Reducing Sea Turtle Bycatch with "TEDs"

Shrimp fishers in the Atlantic and the Gulf of Mexico are now required by law to use ***Turtle Excluder Devices (TEDs)*** on their nets. TEDs are grates (openings with bars across them) located in the cod-end of trawler nets. They allow large organisms such as sea turtles and big fish to escape from the nets without being hurt. Until TEDs were required on shrimp boats, up to 44,000 sea turtles were killed in the Gulf every year. TEDs have helped reduce that number, but even *with* their use, many sea turtles continue to die. Six out of seven species of sea turtle are now in danger of extinction.

Reducing Other Bycatch with "BRDs"

Even with TEDs, bycatch of other, smaller animals is still a problem. One way to reduce this smaller bycatch is with ***Bycatch Reduction Devices (BRDs***, or "birds"), which direct fish through escape "windows" in the trawls. The Gulf of Mexico Regional Fishery Management Council now requires the use of BRDs on all shrimp boats.

Damage to the Ocean Floor

As bottom trawlers drag their nets along the ocean floor they destroy reef-building animals, sea grasses, sheltering seaweed, and other organisms that may provide important habitat or cover for many species of young fish. Trawling also stirs up the bottom and makes the water cloudy, which may block light needed by algae growing on the sea floor and limit the variety of bottom-dwelling organisms.

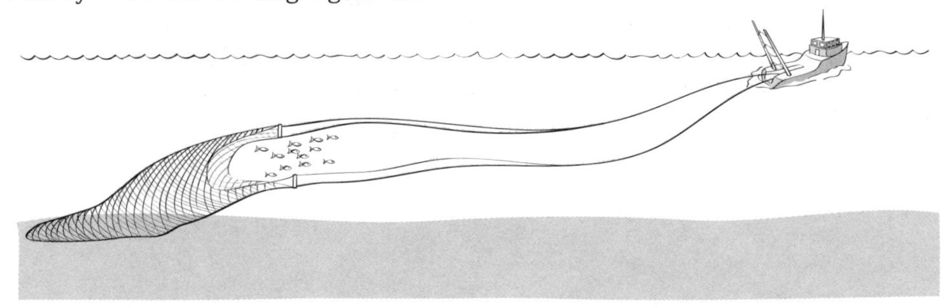

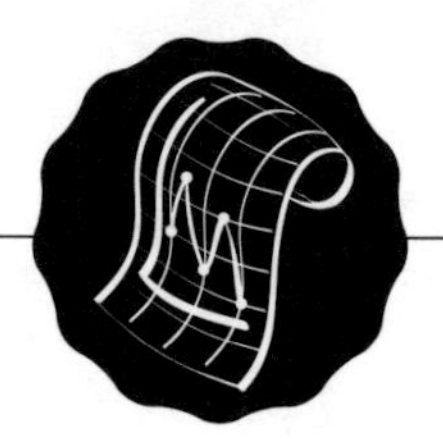

What's Happening with the Fishery?

A Big Cost to the Environment

Shrimp fishing is big business...but shrimp fishing is hard on the environment. One-third of the world's 30 million metric tons of ***bycatch*** (unwanted organisms thrown back into the ocean) comes from shrimp fishing. For every pound of shrimp caught by trawlers in the Gulf of Mexico and southern Atlantic, up to 10 pounds of something else—fish, sea turtles, and invertebrates—are caught along with them and shoveled back into the sea, dead or dying. The number of shrimp boats in the Gulf has increased dramatically in the last 20 years. Because of all the crowding, trawlers are often pushed into fishing grounds that aren't very productive for shrimp—but are productive for other species. So this crowding and competition in the shrimp fishery results in increased bycatch of other species.

Reducing Bycatch

To reduce the bycatch of large marine animals such as sea turtles, U.S. shrimp boats are now required to use ***Turtle Excluder Devices (TEDs)***. TEDs allow sea turtles and large fish to escape unharmed through an opening at the end of the ***trawl*** (net). Another bycatch-reduction method, called a Bycatch Reduction Device (BRD, or "bird"), nudges smaller fish through escape "windows" in the trawls, and greatly reduces the bycatch. As more and more trawlers use them, BRDs should reduce fish bycatch 40–50 percent.

How TEDs Affect U.S. Shrimp Fishers

For many years, countries that allowed their shrimpers to fish without using TEDs had an advantage in the U.S. marketplace. Since those fishers didn't have to spend money buying and maintaining TEDs, they could afford to sell their shrimp more cheaply than the U.S. shrimpers, who *did* have to use TEDs. This made other countries' shrimp more popular than U.S. shrimp. In the mid-1990s, however, the U.S. government changed that: it banned imports of shrimp from countries that don't require TEDs on their shrimp trawlers. Now that all fishers have to use the same equipment to sell shrimp in the U.S., everyone has a fair chance in the marketplace.

Setting Limits

In 1999, the ***maximum sustainable yield (MSY)*** for shrimp (the amount of fishing that can occur while still leaving enough animals to reproduce) was about 120,000 metric tons. To make sure enough shrimp survive to sustain the fishery, U.S. shrimpers are only allowed to fish at certain times of the year. These times depend on the annual life cycles and growth rates of the shrimp. When 50 percent of the brown shrimp in coastal waters are big enough so that a pound contains no more than 100 whole shrimp, the spring shrimping season opens.

Shrimp "Farming"

With almost all the world's wild shrimp either fully fished or overharvested, and the worldwide shrimp fishery in crisis, the shrimp industry has turned to underwater coastal "farming," or aquaculture. Aquaculture now provides more than one-fourth of the shrimp sold and eaten around the world. That figure is expected to climb to 50 percent or more as the wild fishery is overharvested and shrimp farming expands.

The Trouble with Aquaculture

Most shrimp farms are located in Asia and Latin America. Building these underwater farms (artificial environments) destroys the coastal habitats needed by many marine organisms. In addition, farmed shrimp are fed "fishmeal," a ground-up or pellet product made from fish, similar to shrimp's natural food in the wild. As more and more shrimp farms need more and more fishmeal, the fishing pressure increases across the planet. (One-third of all fish caught from the ocean are being ground into fishmeal and other animal feed every year.)

(continued→)

Managing the Fishery

Laws in many countries (Australia's a good example) now regulate their shrimp fisheries by:

- limiting how many trawls may be pulled by one boat;
- limiting how big the nets and boats may be;
- limiting the length of time the nets may be pulled across the ocean at one time; and
- requiring commercial shrimpers to have both TEDs and BRDs on every shrimp net.

Close enforcement of these regulations is critical to their success.

The table below shows in metric tons (mt) the total **shrimp** caught by all methods by all countries of the world.

Year	Weight
1980	1,677,652
1981	1,622,210
1982	1,742,742
1983	1,824,362
1984	1,938,498
1985	2,166,159
1986	2,292,308
1987	2,427,121
1988	2,566,164
1989	2,569,388
1990	2,629,041
1991	2,876,374
1992	2,991,277
1993	2,917,343
1994	3,133,328
1995	3,236,793
1996	3,362,673
1997	3,537,458
1998	3,764,934

What's the Big Deal?

Who Depends on Fish

Of the 30 countries most dependent on fish as a protein source, 26 are in the "developing world." People in developing countries such as Haiti and Lesotho depend more on fish as their main source of protein than people in industrial countries such as the U.S. and Japan. But people in industrial countries eat 40 percent of the world's fish!

How Much We Eat

Between 1995 and 1997, people in the U.S. ate an average of 46 pounds of fish per person ***(per capita)*** every year. The Japanese, biggest harvesters of fish in the world, ate a hefty 152 pounds per capita. Spain consumed 90 pounds per capita, per year, while people in Afghanistan and Ethiopia averaged only .2 pounds per person.

What Else Gets Caught

The total worldwide catch of all species of fish and shellfish is about 93 million tons. Approximately one-third of that catch is wasted—thrown back into the sea dead or dying. That's around 30 million tons—as heavy as 525 average-sized cruise ships! These unwanted marine organisms are called ***bycatch***, or incidental take.

Shrinking World Fisheries

From 1950 to 1989, there was a 300-percent increase in the amount of marine fish caught by the world fisheries. Since then, the total catch has been *decreasing* due to overfishing. Nearly every individual fishery is in decline.

#1. What's Known about It?

1. On poster paper, create a title for your poster with the name of your fishery species ("target species") and the name of your assignment. Be sure to write your own name underneath.

2. Look at the provided illustration of your fishery species. On your poster, make a detailed and scale drawing of your organism. Include a **scale** to represent its actual size in real life—for example, how many feet are represented by one inch on your drawing. Use color to show how the organism looks in real life.

⊢⊣ 1 inch = 1 foot

3. Looking again at the provided illustration, what characteristics do you think could be used to "key out" (identify) this species? Label those parts on your own drawing.

4. After reading your Fishery Information Card, use your own words to describe three or four things about the biology and life cycle of this species. (Note: What's not known can be as important as what we do know!) Place this description near your illustration.

5. Pretend that, as a fisheries biologist, you want to persuade the National Marine Fisheries Service (NMFS) to give you money to support your research. What else do you think needs to be researched about this organism? (This could be something you were unable to learn from the Fishery Information Card or marine reference books, or something "extra" you're interested in about this species). On your poster, write a paragraph to NMFS about what you're attempting to find out and why your research is so important. Begin your paragraph, "What I want to research about ________*[insert your fishery species]*________ is…."

#2. What's for Dinner?

1. On poster paper, create a title for your poster with the name of your fishery species ("target species") and the name of your assignment. Be sure to write your own name underneath.

2. On your poster, list the predators and prey of your fishery species.

3. Looking at the provided illustrations, make detailed, **scale** drawings of one predator and one prey of your target species. Include a scale to represent their actual size—for example, how many feet are represented by one inch on your drawing. Use color to show how the predator and prey look in real life.

⊢⊣ 1 inch = 1 foot

4. **If your fishery species is SWORDFISH, TUNA, WALLEYE POLLOCK, OR SQUID, create the following graph:**
Make a circle graph to show the fractional part, or percentage (%), of each prey item in the overall diet of your target species. Label your graph "Typical Diet of ___*[insert your fishery species]*_____," and label each fractional part.

If your fishery species is SHRIMP, create the following graph:
Make a circle graph to show the fractional part, or percentage (%), your target species represents in the overall diet of one of its predators. Label your graph "Percentage of Shrimp and Other Prey in Typical Diet of ___*[insert name of predator]*_____," and label each fractional part.

5. If your fishery species population crashed, what might happen to its predators and prey? Make a multiple-line graph showing the possible outcomes. (See the note below and be ready to explain why you created your graph a certain way.) Show one line representing your target species, one line representing the predator(s), and one line representing the prey. Label the *y*-axis "population size" (indicate what units of measure you're using) and the *x*-axis "time in years." Label your whole graph "Possible Collapse."

There are many possible predictions. Here are a couple:

1. If the target species crashed, its predator(s) might not have enough to eat. The predator population might also crash. The target species' prey, on the other hand, might have a population explosion *once the target species crashes and stops hunting it.*

OR

2. Removing the target species might not *affect the predator—the predator might just start hunting a different species. The prey might also be unaffected—a different hunter species might replace the species that crashed.*

What are some other possibilities?

#3. Where in the World?

1. Paste the attached World Map Handout on a piece of poster paper, near the top. Create a title for your poster with the name of your fishery species ("target species") and the name of your assignment. Be sure to write your own name underneath.

2. If your map isn't already labeled, locate and write in the names of the countries and ocean basins mentioned on your Fishery Information Card. Referring to the appropriate chart on your card, color your map to show **where *most* of the fishing for your species** takes place and where the *least* amount takes place. (Use any colors except red and blue, which you'll use for something else.) Be sure to include a legend (key) explaining what each color represents.

3. Referring to another chart on your Fishery Information Card, find the country that **catches the *most* of your species.** Draw on your world map four fish for the country that catches the *most* of your species. Then draw one fish for the country that catches the *least.* Finally, pick two countries that fall in between, and label one with three fish and one with two fish. Add a legend on your map showing what each number of fish symbols represents.

4. Referring to the third chart on your card, show on your world map who **eats most of the world fishery catch of all fish and shellfish species:** Draw four forks for the country that eats *most* of the fish caught, and one fork for the country that eats the *least.* Pick two countries that fall in between, and label one with three forks and one with two forks. Add a legend on your map showing what each number of fork symbols represents.

5. Color the major ocean current systems in the area where **your fishery** is located, as marked on your map. Use red for warm-water currents and blue for cold currents. Think back to the activity Apples and Oceans. Why do you think the fishery is located where it is? Write your answer on your poster.

#4. How Are They Caught?

1. On poster paper, create a title for your poster with the name of your fishery species ("target species") and the name of your assignment. Be sure to write your own name underneath.

2. Look at the provided illustration of the fishing gear (equipment) used in this fishery. On your poster, make a detailed and **scale** drawing of the gear. Include a scale to represent its actual size in real life—for example, how many feet are represented by one inch on your drawing.

1 inch = 1 foot

3. Referring to the descriptions on your Fishery Information Card, label your drawing with the important parts of the gear used in your fishery.

4. Use your own words to write three or four sentences describing how this fishing method works. Place this description near your drawing of the gear.

5. What, if any, are the problems with this fishing method? Draw a picture of one or two **bycatch** organisms, if you can, or illustrate the kinds of changes to the habitat caused by this type of fishing. Be sure to include a title for your drawing; something like "The Problems with _______*[insert the fishing method for your species]*_________."

6. Describe on your poster how some people (including scientists, environmentalists, or fishers) would like to change or manage this fishing method in order to:

- decrease the bycatch
- decrease the possibility of overfishing
- decrease the habitat destruction it causes

Write a title at the top of your description; something like "Proposed Changes for the _______*[insert your species name]* _______Fishery."

#5. What's Happening with the Fishery?

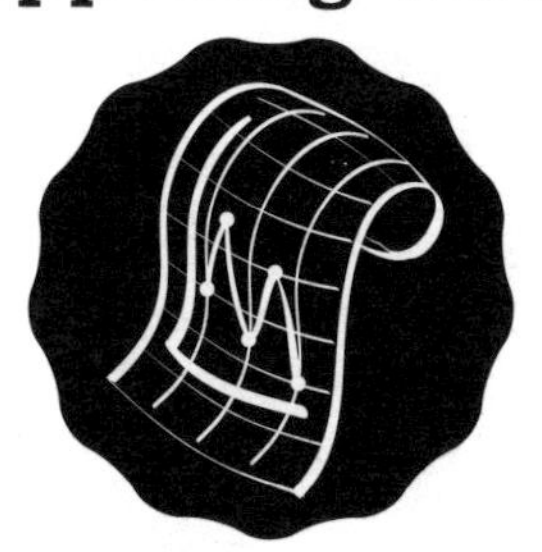

1. On poster paper, create a title for your poster with the name of your fishery species ("target species") and the name of your assignment. Be sure to write your own name underneath.

2. On your poster, create a line graph to represent the ***landings*** (how much of this type of fish was caught and brought to shore) as shown on the table in your Fishery Information Card. Label the *y*-axis "landings," using metric tons (mt), and label the *x*-axis "year caught."

3. The ***maximum sustainable yield (MSY)*** represents the most fish that can be taken and still leave enough to reproduce, so that the population doesn't continually get smaller. If it's known, add the maximum sustainable yield to your graph and label it MSY. If the MSY is not known, write "maximum sustainable yield (MSY) unknown."

4. Label your completed graph. Make sure you indicate in the title of your graph whether the landings are global (from all countries that fish for this species) or from only one country. Does this fishery appear to be regulated in a sustainable way? Would you describe the fishery as being well managed or overfished? Using a title such as "______*[insert your fishery species]*___ Fishery Management," describe on your poster how well or poorly this fishery is managed at this time.

5. Describe two things mentioned on your Fishery Information Card that can affect how many fish are caught in a year. Point out some of the yearly variations and trends as shown on your graph. Again, include a title for this description.

#6. What's the Big Deal?

1. On poster paper, create a title for your poster with the name of your fishery species ("target species") and the name of your assignment. Be sure to write your own name underneath.

2. Carefully read the statements on your Fishery Information Card, and consider at least three points about fisheries you think are especially important.

3. On your poster, make at least two types of graph to represent the information you've chosen to focus on. Be sure to give a title to each graph and to label all your data.

4. Now, represent your important fishery facts in **ONE** of the following ways on your poster. Be sure to create a title for whichever option you choose.

a. Draw a picture (or a series of pictures) showing three new things you learned about fisheries and would like to share with others.

OR

b. Create an editorial cartoon, like the example provided, to represent two or three fishery messages you think are most important. Use both images and words.

OR

c. Use poetry (either one long poem or several *haiku*—see note below) to express three or more sentiments you'd like to share about the ocean's fisheries.

Note: Haiku *(pronounced "hy-koo") is a form of poetry created in Japan several centuries ago.* Haiku *(that's the plural, too) are very short, expressive poems, typically written in the present tense. They're only three lines long—traditionally with five syllables in the first line, seven in the second, and five again in the third. Unlike much western poetry,* haiku *do not rhyme. Each of your* haiku *should be three lines long, non-rhyming, and written in the present tense. You may use any number of syllables you like per line. See the examples provided.*

What are they catching,
The small boats in the offing,
As snow falls on my *kasa*?
—Kuroyanagi Shōha (d. 1772)

One cherry blossom
beautiful as it can be
the aspect of life.
—Kalynda Stone, Grade 7

One field of frogs
croaks for a time,
And then is silent.
—Mukai Kyorai (1651–1704)

OR

d. Write a composition, about three paragraphs long, that might persuade someone to take an interest in the condition of the world's fisheries.

But I didn't order all that!
Oh, but it comes with all this—at no extra cost (to you)!

WORLD MAP

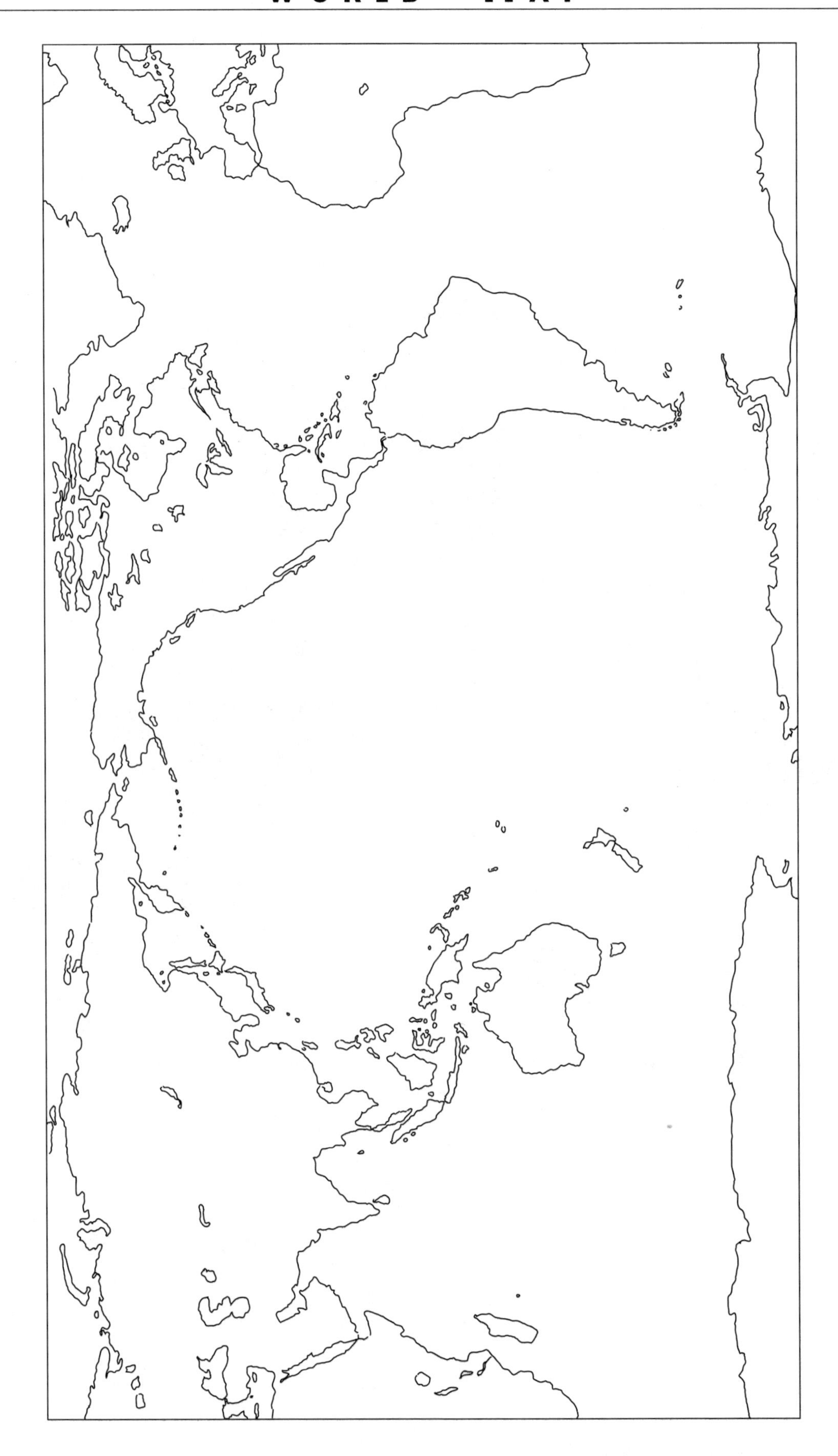

Behind the Scenes

The following information, organized according to the three main activities in this guide, is provided as background for the teacher. This section isn't meant to be read aloud or distributed directly to your students. It's intended to provide necessary and concise background for you in presenting the activities and responding to student questions. Please see "Resources" on page 163 and "Literature Connections" on page 176 for books and other materials that may help you and your students delve more deeply into the interconnectedness of our ocean.

Activity 1: Apples and Oceans

Most of our planet is covered by ocean. The Southern Hemisphere, with only one-third of the land area on Earth, could easily be called the Ocean Hemisphere. The ocean is a major distinguishing feature of Earth, making our planet different from all others in the known solar system. It's impossible to understand the biological, geological, or human history of our planet without first understanding a bit about how the ocean works.

We now know that the "seven seas" of maritime lore are a romantic relic. In truth, these "seas" are ocean basins within one large expanse of ocean. In our modern world—where the consequences of our actions are often global—what's put into one "sea" may very well end up on the shores of another, halfway around the world. It's more important than ever to realize that there's "only one ocean."

Here's one example of how ocean currents affect the entire planet. We know that trash from the coasts of North America regularly washes up on the remote South Pacific island of Laysan, in northwestern Hawaii. After every breeding season of the Laysan albatross, small piles of plastic are found in their nests—evidence that the parent birds collected it out at sea, thinking it was squids or plankton, and fed it to their chicks. Every year, chicks die of starvation, decompose, and leave behind only their stomach contents...inedible, non-degradable bits of plastic. These, in turn, blow back out to sea to be collected by future generations of birds and sea turtles.

Connected by surface and bottom currents, acting in many respects like the circulatory system of our bodies, sea water—with all its marine organisms and debris—is carried throughout the ocean. Take a walk on the beach in Port Aransas, Texas, for instance: Because of the quirky currents leading into but not out of the Gulf of Mexico, you can regularly find on this town's beaches trash from all over Europe, Asia, and Central America.

Suppose you took a cup of seawater from off the coast of California, another from inside the Mediterranean Sea, and yet another off South Africa's Cape of Good Hope, and let the water in each cup evaporate. You might have different total amounts of salts left behind in each sample—but the constituent salts themselves, and their exact proportions to one another, would be exactly the same! The currents ensure it; there's only one ocean. The composition (makeup) of seawater is the same in all parts of the one world ocean. Equally exciting, the chemical composition of seawater is vitally similar to the chemical makeup of the body fluids of all organisms. Most of the more abundant elements in seawater are also important components of us all.

Although ocean resources seem infinite, there are huge areas of open ocean that support very little life. Almost all the depths below the photic zone (the surface waters through which light penetrates) are considered biological deserts—that is, they support very little life.

Cold, salty, nutrient-super-rich water melts off the ice in Antarctica in summer, sinks to the bottom, and is pushed along 10,000 miles of (uncharacteristically flat) ocean floor all the way to Newfoundland, Canada—where it finally upwells around the steep sides of the Grand Banks. There, until recently, it created one of the most productive and lucrative localized fishing areas in the world—until the area was overharvested and the habitat degraded. The problems associated with overharvesting are addressed in Activity 3: What's the Catch?

Only about 8–10 percent of the ocean (over parts of the continental shelf) supports large concentrations of phytoplankton (microscopic algae), the base of the food web. Worldwide, over 90 percent of the human fishing effort is concentrated in these coastal (neritic) zones along the margins of the continents. Within these coastal zones is a small handful of intensely productive areas, where deep, cold, nutrient-rich waters are upwelled, or brought up to the surface. Once at the surface, these nutrients act as fertilizer to stimulate blooms of phytoplankton living in the sunlit surface waters. The depth to which light penetrates varies from less than one meter in turbid (cloudy) estuaries, where rivers meet the ocean, to over 100 meters in clear, open-ocean waters. This thin surface layer, where photosynthesis occurs, is called the photic zone.

Major areas of upwelling support intense seasonal blooms of phytoplankton off the west coasts of North America, South America, Africa, and Australia. Upwelling also occurs off the coasts of Spain and Portugal, and around Antarctica. Phytoplankton form the base of the food pyramid, and are in turn grazed by zooplankton (microscopic animal life)...which are eaten by fish, which are eaten by hundreds of thousands of seabirds, marine mammals, and larger fish such a sharks. These biologically productive areas also support vast forests of kelp (large, brown seaweed), itself a major habitat for marine organisms. With all the marine life they support, these upwelling areas also attract most of the world's fisheries. Astonishingly, the rich upwelling areas of the world represent only 1/1,000 (.01 percent) of the ocean—only 3/4,000 of the whole Earth. (For more information on upwelling, see page 32 in Activity 1.)

Activity 2: Squids—Outside and Inside

Our knowledge of squids has tentacles reaching far back in history. Squids were known even to Aristotle, and illustrations were made as long ago as 4,000 years. Sailors described squids in their yarns and legends, most often as monsters of the sea, or "kraken." Jules Verne wrote about a fictional, highly exaggerated giant squid that "captured" the submarine *Nautilus* in *20,000 Leagues Under the Sea.*

Modern biologists don't know much more about giant squids *(Architeuthis dux)* than researchers did in the 1800s, because most giant squids available for study have washed ashore in pieces, rather than as whole animals. Nobody knows how big they get; the largest one ever found was from the stomach of a sperm whale, and it measured 65 feet. Scientists speculate that they may grow up to 100 feet long and weigh more than two tons. Giant squids live at depths of 1,000 to 3,000 feet; it's no wonder they're difficult to study!

Squids don't have to be giant to be excellent predators. The most ferocious is the jumbo, or Humboldt, squid, which lives off the coast of South America and reaches 12 feet long and 300 pounds. These squids travel in shoals and attack their prey *en masse*.

Squids are distributed worldwide and come in many sizes and shapes, from less than one inch to greater than 70 feet. They have no backbone, and are therefore classified as invertebrates. Giant squids are the largest invertebrates in the world. Squids are related to the octopus, cuttlefish, clam, and snail—all classified as phylum Mollusca. Squid characteristics found in other mollusks include the soft, bilaterally symmetrical body; the foot (here modified into greatly expanded tentacles); the mantle that surrounds the organs of the body; the ribbon-like, raspy tongue called the radula (missing in the clams); and a remnant of shell (greatly reduced or missing in some other mollusks too). Squids are in the class Cephalopoda, meaning "head-foot," along with octopuses and cuttlefish. There are about 400 living and about 10,000 fossil species of squids in the world.

Cephalopods have the most highly developed nervous systems of all invertebrates, and the most complex behavior. Their sense organs, brain, and excellent swimming ability allow them to compete with vertebrates for food and territory in the same habitat. Squids are adapted to a completely aquatic way of life, with gills, a streamlined body, a reduced shell, and a funnel for speedy movement. Specialized adaptations allow them to be camouflaged and yet move freely and fast through the water under their own power. Like fish and a wealth of other, diverse organisms (including whales), the squid is considered a pelagic animal, meaning it spends its life in the open ocean.

Scientists are studying the squid's ability to learn and remember, and how its tentacles are used for touching and identifying different objects. Researchers are also studying the squid's giant ***axon,*** the long, string-like extension by which a nerve sends its electrical signals. These signals tell muscles to contract and glands to secrete hormones, and send information from the sensory organs to the brain. A squid's axons are each about the width of a pin—up to 100 times larger than in humans. The squid's jet-propulsion system lets it move at blazing speeds; this method of locomotion explains why the nerves that activate the muscles have evolved such big axons. The axons' size makes them ideal for study, and much has been learned about the human nervous system from studying the squid. (See also the sidebar under brain dissection.)

A field biologist we know phoned from the Channel Islands one night during production of this guide to report that he was reading *just by the lights of the squid boats offshore. One of the boats chugged past him in the dark—looking for all the world, he said, "like a floating, lighted soccer stadium."*

Pacific Coast market squids *(Loligo opalescens)* are captured in large numbers off the coast of California as they form into mating shoals. There are two main fisheries for these squids in California: Monterey Bay, where they're caught during the summer months, and the Channel Islands in Southern California, where they're caught in winter. In the U.S., 90 percent of squid fishers use special nets called purse seines to make their catch. Fishers shine large, bright lights over the water at night to attract spawning squids to their boats, then haul in the nets. A different species of squid is caught by Japan, which takes 80 percent of the world's catch.

Squids are eaten by many different predators and are a vital food source throughout the world's oceans. Sharks, pilot whales, porpoises, sperm whales, elephant seals, sea lions, sea otters, salmon, swordfish, and such marine birds as auks, penguins, and terns all depend to different extents on the squid for food.

Squids are also a very important food resource for people around the world This in turn makes them an excellent dissection animal; dissecting something that can be eaten afterwards discourages a casual attitude toward taking an organism's life. We need to make clear to students that we're not simply going to throw away this once-living being after we're done with it, and that we're not just experimenting for experiment's sake.

Activity 3: What's the Catch?

We may destroy the land and its resources with too many people and little thought for the future. We may pollute and overwork the land with methods that diminish its capability to replenish nutrients. But the *ocean*...well we've always taken for granted that we could turn to the ocean and its resources indefinitely. But can we? Can we still count on the ocean to feed the endlessly growing world population once we've exhausted the land and *its* resources? It's predicted that as the world population increases, the demand for fish will climb to about 120 million tons by the year 2010. At the rate we're going now—fishing and overfishing—we're looking at a 20 million–ton shortfall, at least, of fish to feed the world.

From 1950 to 1958, the world's catch of fish from the oceans steadily increased from about 20 million tons to 90 million tons every year—that's a 300-percent increase in the global marine fish catch. The catch continued to rise for over 30 more years; then, the bottom dropped out of this seemingly inexhaustible catch of protein. In the first three years of the 1990s, for the first time in history, the catch dropped. Money was thrown at the fishing industry to buy more fishing boats equipped with the most up-to-date technology. The catching power of the world's fishing fleet increased tremendously... and still the catch stopped growing. What was happening? What's affecting our fisheries to this day?

One of the major causes of the continued decline is simply this: too many boats catching too many fish. Most of the world's marine fisheries are at or beyond the limits of sustainability. This means that so many fish are being caught—in some cases, 90 percent of the fishery population—that they can't be fished commercially and still maintain a large enough population to reproduce themselves. They're being fished beyond the maximum sustainable limit. By 1997, according to the United Nations, all but two of the 15 major ocean fishing grounds had been overfished (overharvested). Four had actually been completely "fished out." The same is true today—and more fisheries are under threat.

A few hundred years ago, the maritime territorial limit of a nation (the waters along its coasts that it could claim and defend) were established at three miles from shore. That was the length of a marine league—not so coincidentally, about the distance a cannonball could travel. Beyond that limit were the high seas, wild and unregulated and open to anyone. The three-mile limit endured until the very recent past.

By the late 1970s, the United States and almost every other nation with an ocean coast had extended "their" waters (the marine areas over which they had jurisdiction) to 200 miles offshore. This 200-mile band is each nation's "Exclusive Economic Zone," or EEZ, and only that country can decide who fishes there and what methods are used. Even now, however, in the United States, over 80 percent of the fisheries within 200 miles of our coasts are overfished—or being fished right at the sustainable limit. And that's just for the fish species for which the status is known; little is known about the condition or status of as many as 34 percent of species fished in U.S. waters. We still have much to learn about our vast ocean, including the effects of our actions and of environmental changes—from global warming to *El Niño*.

Some scientists feel that to sustain marine life as we know it today, we should be preserving a minimum of 20 percent of our EEZ as Marine Protected Areas.

Overfishing of target species (the fish the fisher is after) isn't the only problem. About one-third of the total annual marine catch is taken unintentionally and thrown back—usually dead or dying—because it's not what the fisher wanted. This incidental take, or ***bycatch,*** contributes greatly to overfishing. Many of the wasted fish are either the juveniles of other commercially important species, or commercially valuable themselves. Endangered marine species such as sea turtles, and other protected wildlife such as marine mammals or seabirds, are also often caught and discarded as bycatch. Some fishing methods result in extreme bycatch and habitat destruction. Bottom trawls, especially, as they drag heavy, weighted nets across the ocean floor, alter and destroy sensitive habitats and take everything in their paths—including tremendous amounts of bycatch.

Coastal waters such as estuaries, marshes, and kelp forests are home to the greatest density of fish in the ocean. In these places, where land meets sea, nutrients and shelter create the most fertile conditions anywhere on Earth. More than 75 percent of the commercially important marine fish in the U.S. depend on estuaries and inshore waters at some stage of life. But the ever-increasing number of people living, polluting, and building in coastal areas threatens this habitat. Marine scientists warn that loss of these coastal waters is the most serious threat to marine fisheries in the United States.

The world's human population in 2001 is about 60 billion. That number is expected to grow by over 60 million people per year. *Nearly half these people will live within 100 kilometers (620 miles) of a coastline.*

Fisheries are a critically important source of food for the world. People in "developing" countries such as Haiti or Lesotho rely much more on fish as an essential source of protein than do people in industrial countries, yet the United States is the world's second-largest importer of fish. Many popular imported species come from fisheries that are poorly managed. Many populations of tuna, for example, are overfished. And tons of shrimp are grown on aquaculture farms whose construction destroys coastal habitats.

Over one billion people in Asia now depend on ocean fish for their entire supply of protein. In Africa, the number is one in five.

Most fisheries of the world are open to all comers; anyone with a boat and a net can go out and fish. This free and open access creates many problems—there are few reasons to fish conservatively if someone else will snag the fish you leave behind. Governments often wait to set rules until the catch starts noticeably decreasing—by which time there are already too many boats fishing. Consumers, scientists, environmental activists, and many concerned fishers are finally making themselves heard loud and clear. Since the fish have their limits, the fisheries are finally being limited.

In 1995, the United Nations supported an agreement on the conservation and management of fish that are highly migratory or that cross the boundary between domestic and international waters. These include swordfish, most tuna, some sharks, and marlin. Once this treaty has been ratified (accepted and passed) by 45 nations, it'll begin to help reverse overfishing and depletion of these fish around the world.

The most important law governing fisheries is the Magnuson Act, created in 1976. It created the EEZ mentioned above, and regional councils to manage the fisheries. In 1996 the Magnuson Act was reauthorized and strengthened, and became known as the Sustainable Fisheries Act. This law designates fisheries as "overfished" if they exceed sustainable limits. It also states that habitat destruction is a major threat to fisheries, and that some methods of fishing can be banned in sensitive areas. And the new version of the Act requires, for the first time, that bycatch be reduced. But problems remain, and even the best law is meaningless unless it's enforced—and that takes money. It remains to be seen if this important step can stem the tide of overfishing and allow our fisheries to recover.

At this writing, only about 124 federal enforcement agents and officers ("fish cops," they call themselves) monitor fishery compliance for the whole United States. Working in association with state enforcement counterparts and with the Coast Guard, this tiny arm of the National Marine Fisheries Service (NMFS) is responsible for all near-shore and at-sea fishery compliance.

Several of the fishery assignments in What's the Catch? call for representing data in the form of graphs. Following is a small "refresher" on the uses and formats of various graphs, with examples.

Glossary of Graphs

A REFRESHER ON GRAPHS FOR THE TEACHER

There are many types of graph. Each uses a unique way to represent a relationship among a set of data. In many cases, more than one type of graph can be used to represent the same data. Depending upon how the data is represented, the factual information on the graph can be presented to lead to accurate interpretation of the data, *or* it may be skewed in some way to lead to inferences about the data that are not necessarily true.

The following are examples of graphs used in this guide.

BAR GRAPHS

A **single-bar graph** uses individual bars (either vertically or horizontally) to represent quantities. There are spaces between the bars.

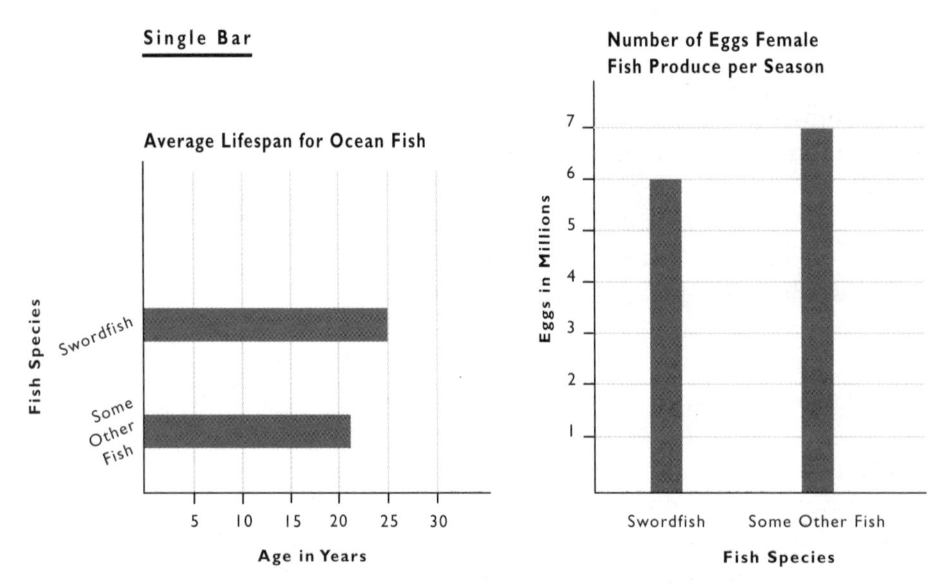

A **double-bar graph** displays quantities in side-by-side pairs, so that comparisons can be shown. The bars are differentiated by contrasting color or shading, and there are spaces between each pair.

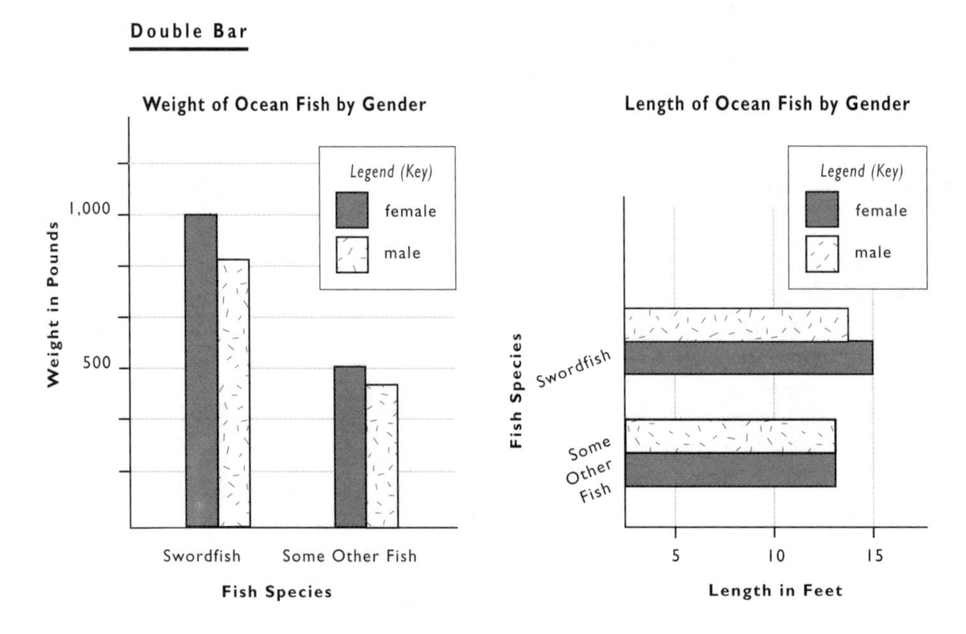

CIRCLE GRAPHS

A **circle graph** displays quantities in relationship to the whole, to show how the size of each part compares to the whole.

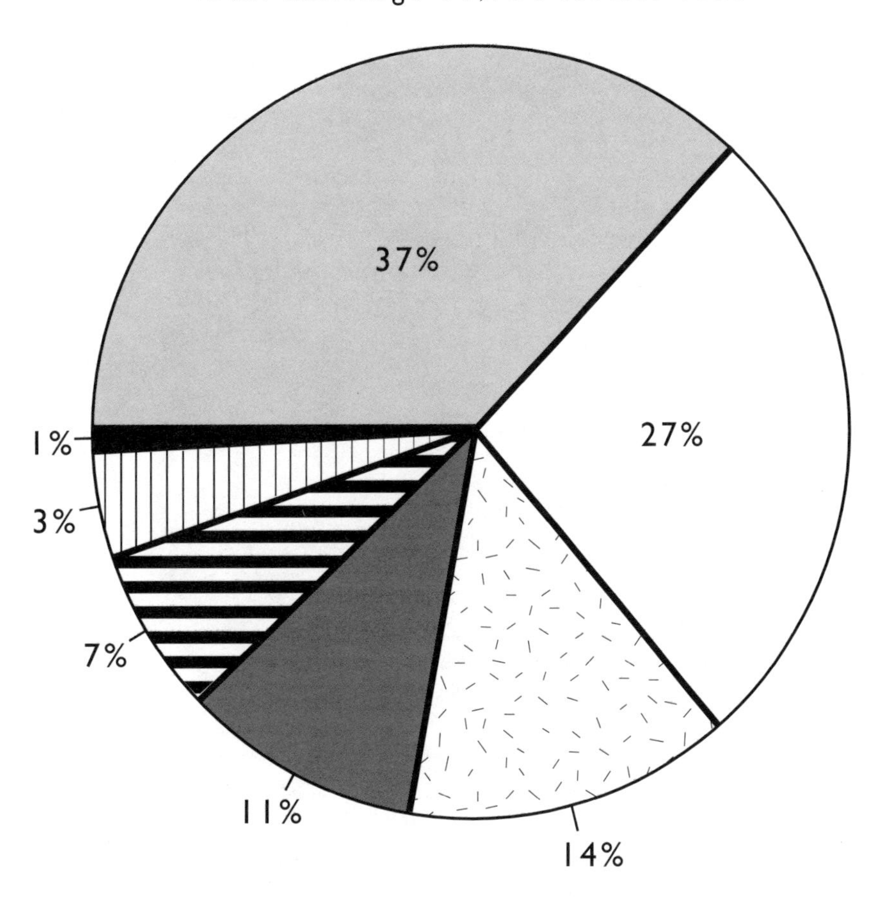

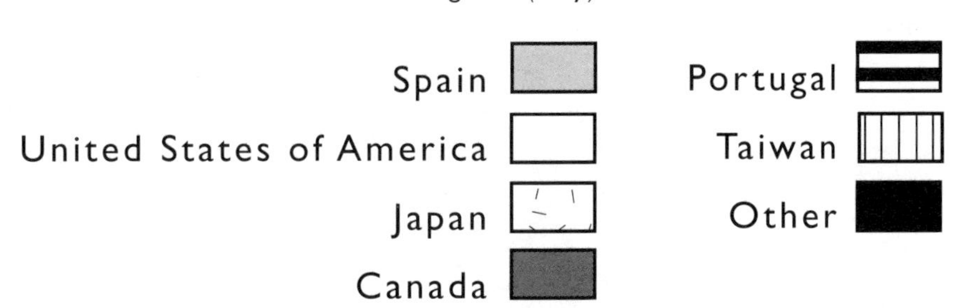

LINE GRAPHS

(NOTE: A line graph is not the graph of a line.)

A **single-line graph** displays change in a quantity over a particular period of time. The line graph consists of a series of plotted points connected by line segments. In the example below, the *x*-axis represents time increments and the *y*-axis represents the quantities.

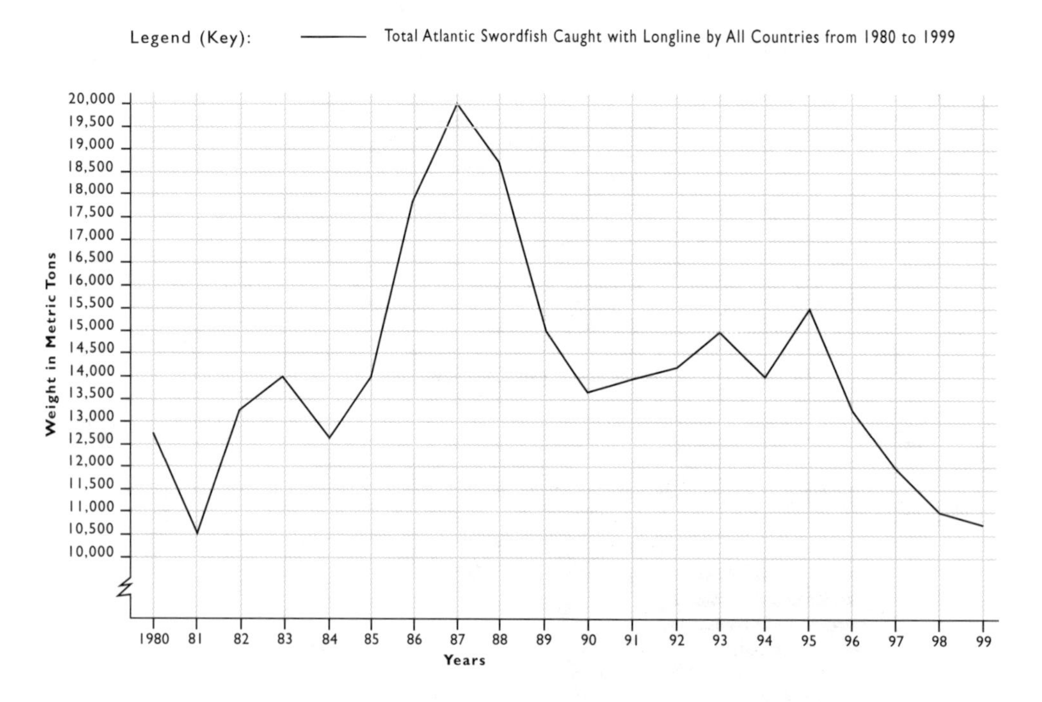

Multiple-line graphs are used to compare two or more quantities that are changing over time. Each line represents one set of data, and usually the lines are carefully distinguished from one another.

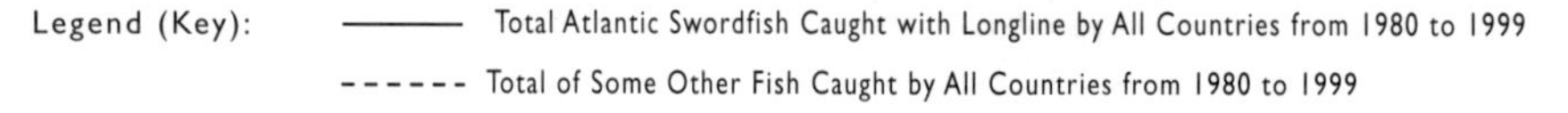

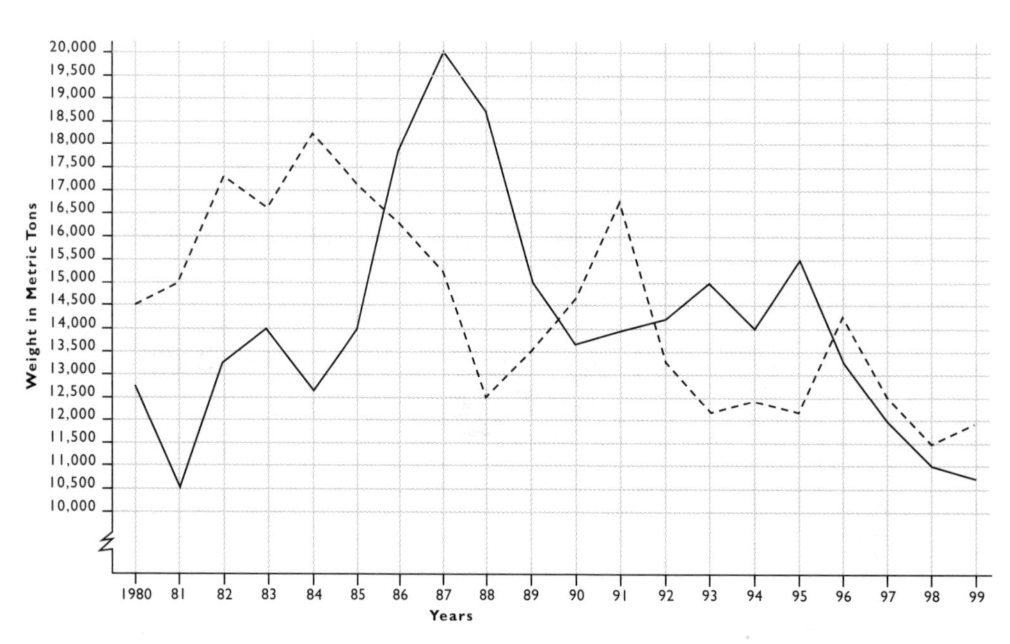

Following are two other types of graphs your students may be familiar with and choose to use in Fishery Experts (Activity 3, Session 2).

PICTOGRAPHS

A **pictograph** uses pictures or symbols to display data.

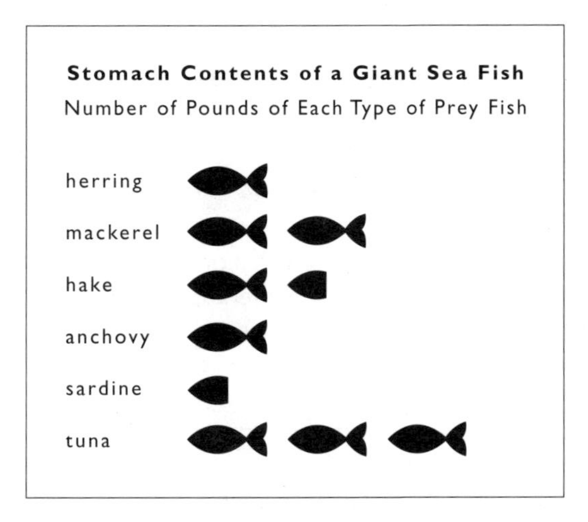

Legend (Key): = 5 pounds

= 2.5 pounds

HISTOGRAM

A **histogram** is similar to a bar graph, except that each bar represents an *interval* and there are no spaces between the bars.

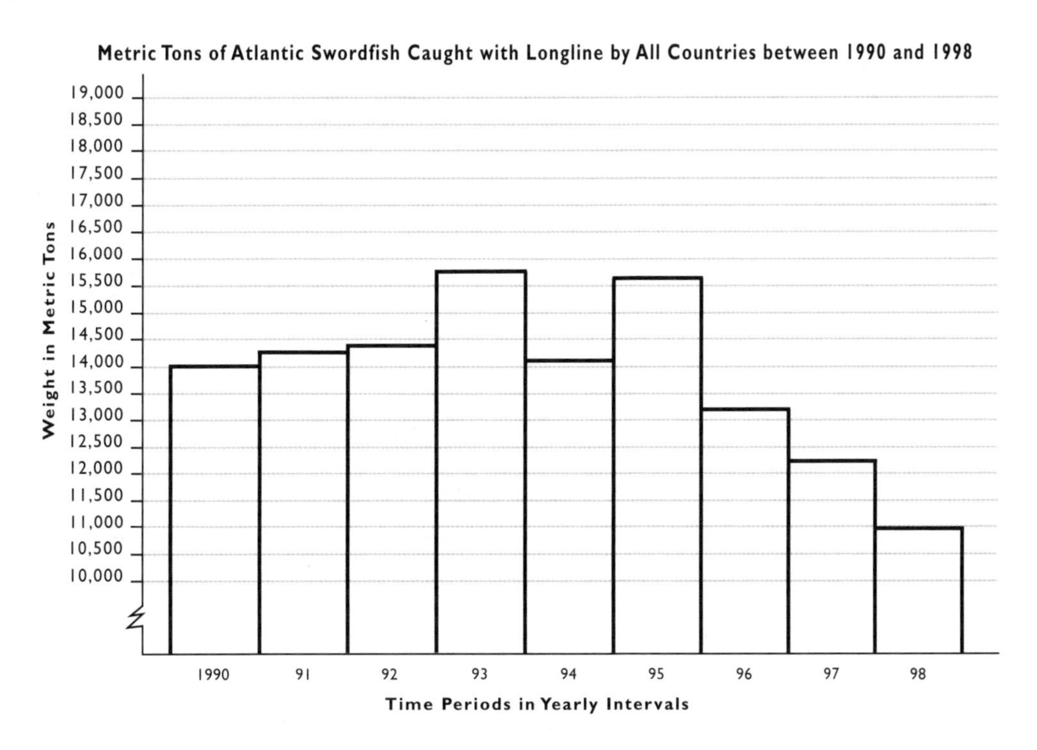

Representing Fractions with Number Lines

Apples and Oceans demonstrates, with an apple model and a circle graph, the fractional proportions of the Earth that are land, ocean, habitable, resource-rich, etc. A **number line** *can be a wonderful additional tool, a useful alternative depiction of fractional parts. Here is a brief review. If the explanation feels technical, a glance at the examples should make it very clear.*

Posting a large model of this number line with the equivalent fractions can serve as a useful tool in conjunction with Apples and Oceans. (Note that this number line shows the fractional parts as a result of successive halving; this is what the activity calls for. Number lines can also be divided into other equal fractional parts, such as thirds, fifths, etc.)

A fraction is a number that represents an equal part of a whole or group. The denominator indicates the number of equal fractional parts in the whole or the set, and the numerator indicates what number of those equal parts the fraction represents. For example, the fraction 3/5 represents the number of fractional parts being described—three—in relationship to the number of fractional parts in the whole—five.

As the denominator gets larger, the fractional parts (the relationship of those parts to the whole or the group) get smaller. This concept is often confusing for students who don't have a conceptual understanding of fractional parts. Fractional parts and equivalent fractions can be made more concrete by using models and representations that clearly illustrate fractional parts.

Making Number Lines

Children often have experience making "fraction strips" and using pizzas or circles as models to understand fractional parts. This provides an understanding of fractional parts of a whole. Building on this, creating a **number line** with your students can provide another, more abstract model that can be referred to when comparing fractions. Since a fraction is a number, a number line can be used to situate fractions and identify equivalent fractions (fractions that name the same number).

The following example connects with the fractional parts in Apples and Oceans. In the case of that activity, the apple and the ocean are the units, or wholes, to be divided into fractional parts.

Begin by creating a number line from 0 to 1. Mid-way between the 0 and the 1 (and one line lower), add the one-half mark. Below the 1, add the two-halves mark. This begins naming equivalent fractions.

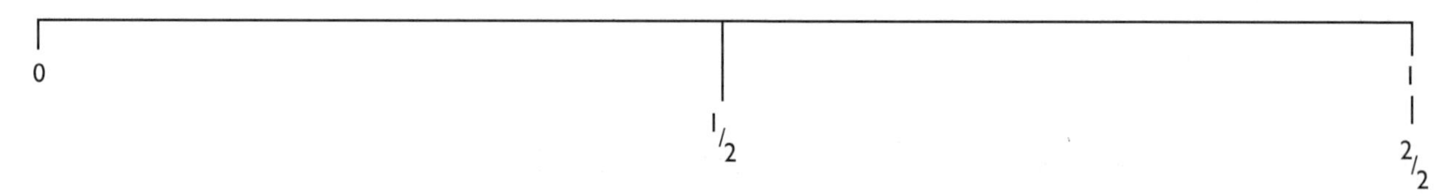

Next, add fourths below the halves to show the following equivalent fractions: 2/4 = 1/2 and 1 = 2/2 = 4/4.

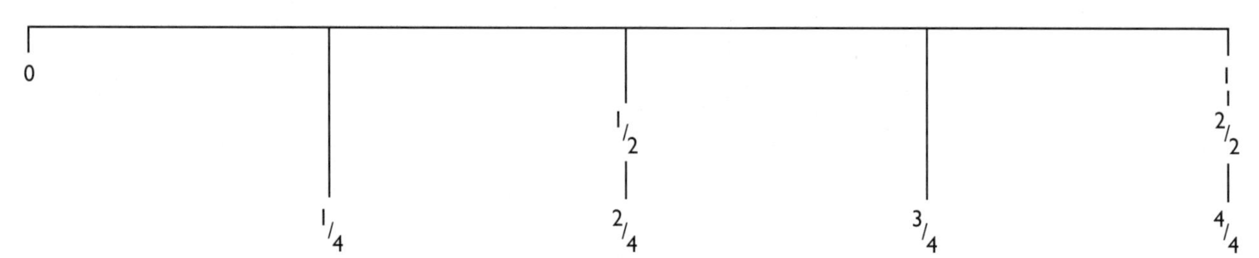

Continue to build the number line by adding eighths. Label all the fractional parts and note the equivalent fractions.

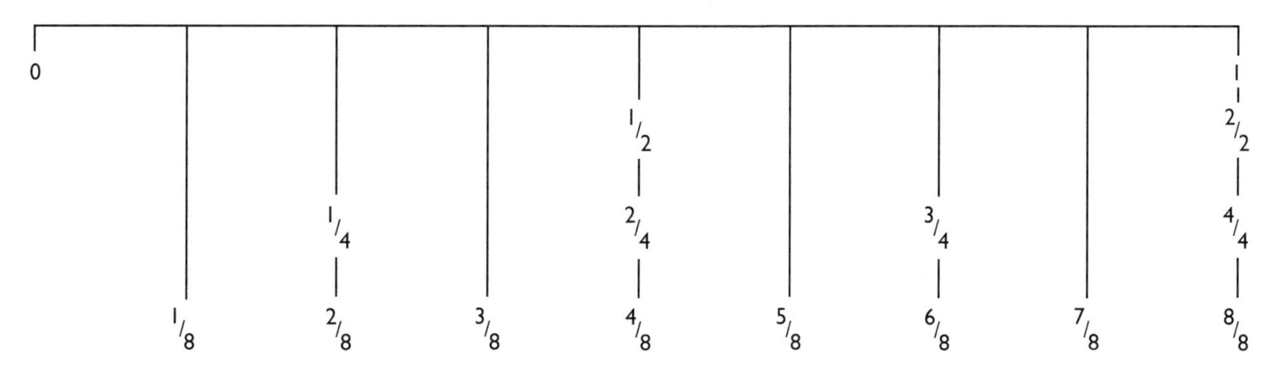

The next fractional parts are sixteenths.

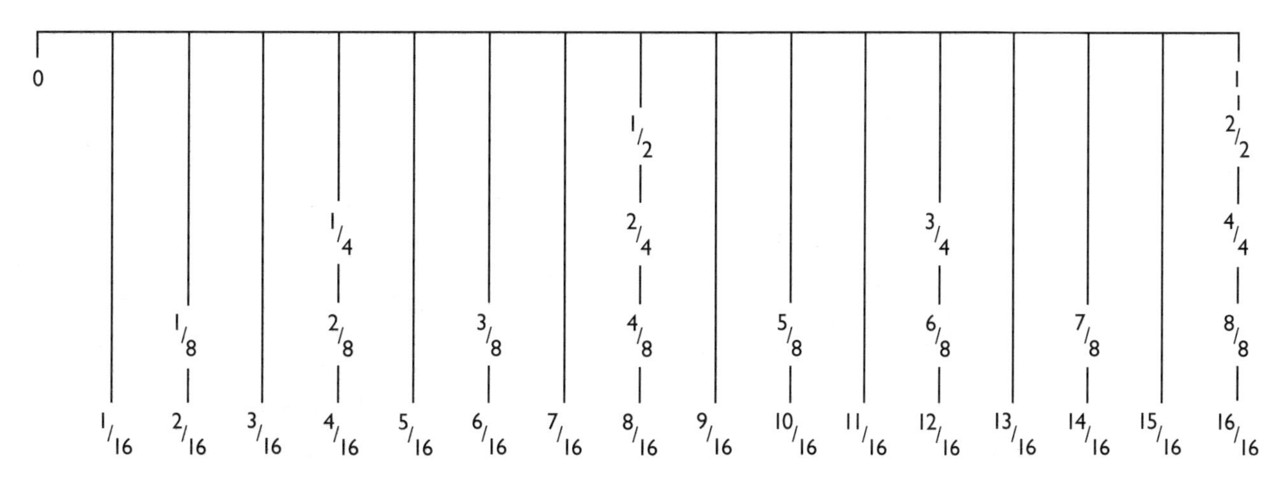

Finally, create thirty-seconds.

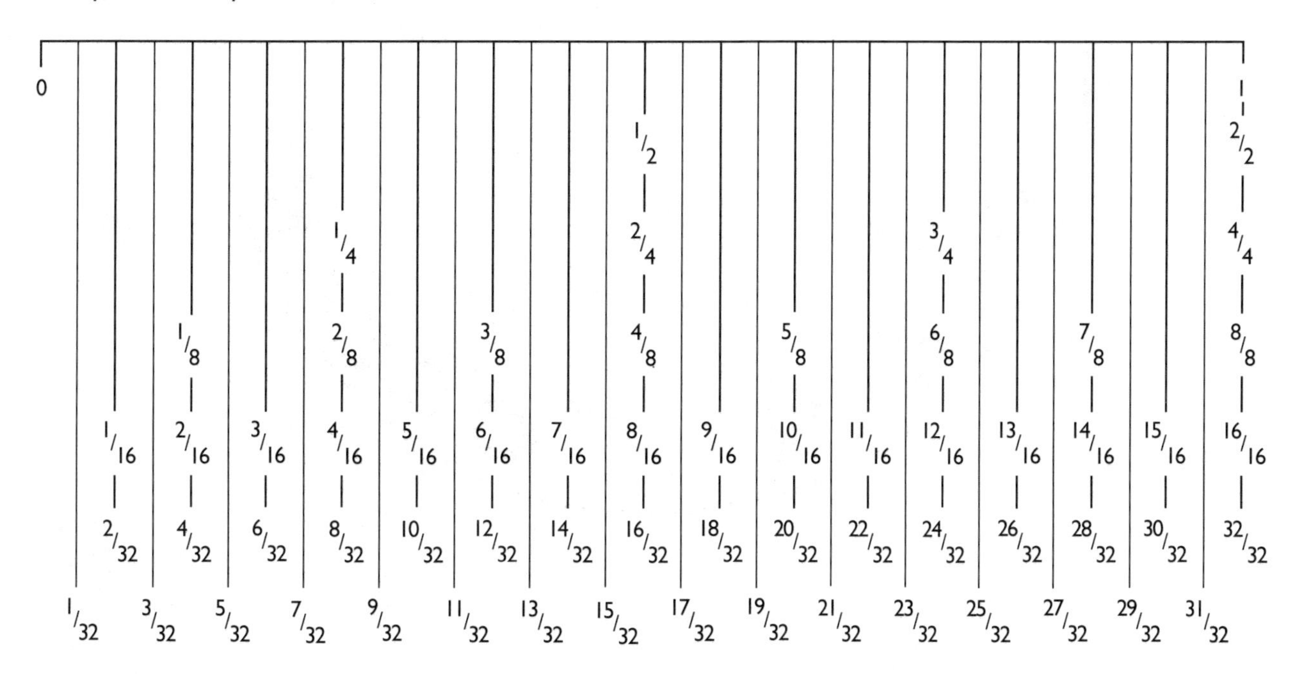

Fractions and Circle Graphs

If we let a circle represent 1 (one whole circle), it can be divided into fractional parts in a way similar to how fractional parts are created on the number line from 0 to 1. The fractional parts into which the circle is divided do not have to be equal to one another, and they do not have to have the same denominator. (However, the sum of all the fractional parts must equal 1.)

A circle graph displays quantities in relationship to the whole, to show how the size of each part compares to the whole. Sometimes the whole is 1, and sometimes the whole represents more than 1.

In Apples and Ocean, the connection is made between the concrete apple pieces and the fractional parts those apple pieces represent on a circle graph. As the pieces are cut into smaller fractional pieces, a representation of each of those fractional parts is made on the circle graph. In that way, there is both a concrete model and an abstract representation. In addition, the students record the fractional parts on their circle graphs as the teacher creates a class record of the fractional parts.

See also page 157: Circle Graphs.

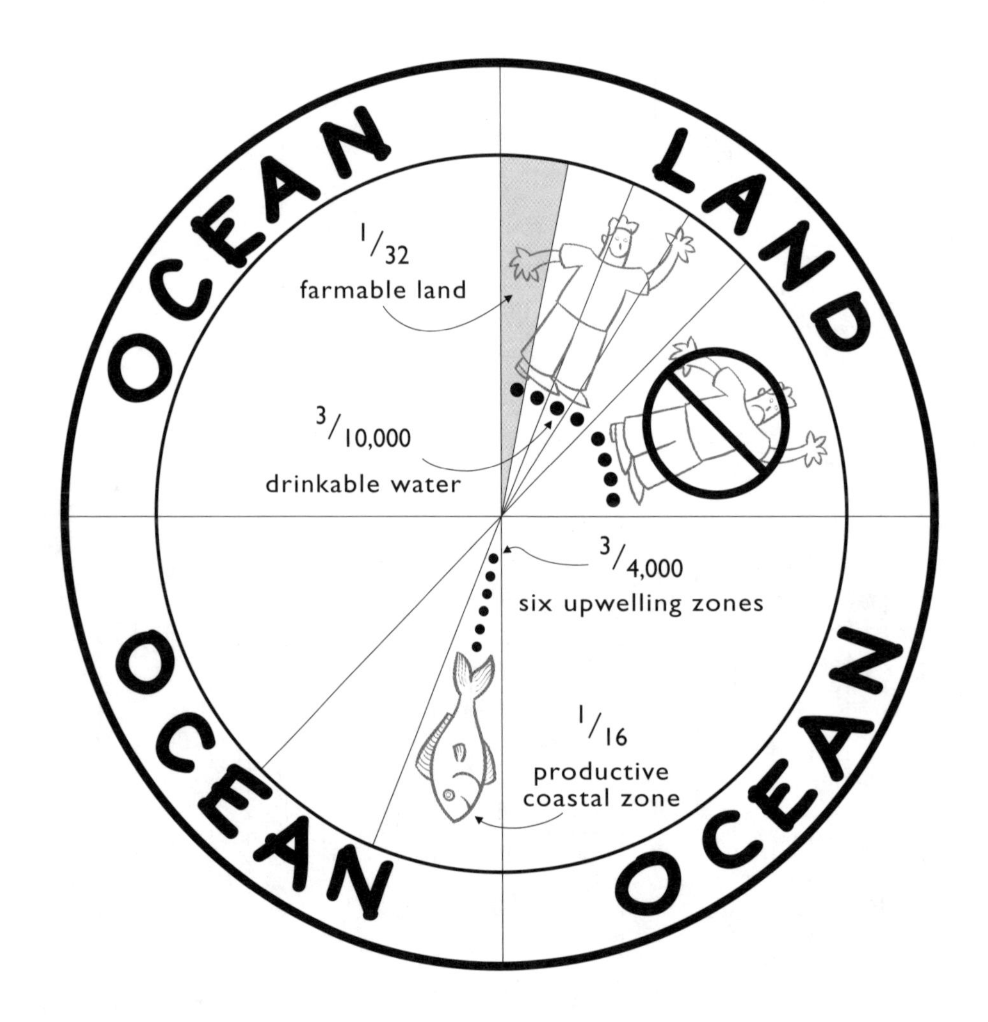

Resources

Please be sure to also see the many excellent books listed in the "Literature Connections" section on page 176.

Books for Students

Bill Nye the Science Guy's Big Blue Ocean
by Bill Nye, with additional writing by Ian G. Saunders;
illustrated by John S. Dykes
Hyperion Books for Children, New York, NY (1999)

Describes the ocean and its life forms, and suggests related activities that explore marine biology.

Kid Heroes of the Environment
Simple Things Real Kids Are Doing to Save the Earth
edited by Catherine Dee;
illustrated by Michele Montez
EarthWorks Press, Berkeley, CA (1991)

Short, catchy profiles of can-do kids and what they're accomplishing, with resources and ideas for up-and-coming activists.

Destination Deep Sea
by Jonathan Grupper
National Geographic Society, Washington, D.C. (2000)

Human use and abuse of ocean resources are the focus of this introduction to a complex topic. From whaling to pollution, land reclamation to desalination, the ocean is viewed in terms of human use.

Faces of Fishing
People, Food and the Sea at the Beginning of the Twenty-First Century
by Bradford Matsen
Monterey Bay Aquarium Press, Monterey, CA (1998)

Highlights the social and cultural changes affecting fishing for future generations. Examines fishing cultures around the world and ways in which fishing is deeply imbedded in community life. Brings readers closer to the global fisheries issues of overfishing, bycatch, coastal habitat destruction, and the impacts of human population growth.

I Wonder Why the Sea is Salty
and Other Questions About the Ocean
by Anita Ganeri; illustrated by Tony Kenyon
Kingfisher, New York, NY (1995)

Nellie Bly's Book: Around the World in 72 Days
by Nellie Bly; edited by Ira Peck
Millbrook Press, Brookfield, CT (1998)

Abridged version of the famous journalist's attempts to circle the world in fewer than 80 days in the late 19th century.

Oceans
by Seymour Simon
Morrow Junior Books, New York, NY (1990)

Portrays the Earth as a water-covered planet on which the continents are islands. Addresses topography of the ocean floor, wave motion, wave energy, and ocean currents.

Oceans
Looking at Beaches and Coral Reefs, Tides and Currents, Sea Mammals and Fish, Seaweeds and Other Ocean Wonders
by Adrienne Mason; illustrated by Elizabeth Gatt; photos by David Denning
Kids Can Press, Buffalo, New York, NY (1997)

Detailed drawings and full-color photos bring the ocean to life. Safe, simple experiments and activities explore the mysteries of the Earth's oceans.

Oceans for Every Kid
Easy Activities That Make Learning Science Fun
by Janice VanCleave
John Wiley & Sons, New York, NY (1996)

An overview of marine sciences, including the techniques and technologies of oceanography, the topology of the ocean floor, movement of the sea, properties of sea water, and life in the ocean.

Octopus and Squid
by James C. Hunt
Monterey Bay Aquarium Foundation, Monterey, CA (1996)

Useful insights and good photos.

Our Oceans
Experiments and Activities in Marine Science
by Paul Fleisher; illustrated by Patricia A. Keeler
Millbrook Press, Brookfield, CT (1995)

An introduction to marine science, with chapters on seawater physics and chemistry, geology (including recent discoveries about the ocean floor), ocean currents and their effect on world weather, and undersea resources and the need to conserve them.

Polluting the Sea (Save Our Earth)
by Tony Hare
Franklin Watts, New York, NY (1991; out of print)

A clear and easy-to-read introduction to a large topic. Discusses various forms of pollution—oil, litter, sewage, metals, chemicals, and radioactivity—and explains why pollution happens, what effects follow, and what can be done about the problem.

Sailing Alone Around the World
by Joshua Slocum; illustrated by Thomas Fogarty and George Varian
Viking Penguin, New York, NY (1999)

Captain Slocum set sail in April 1895 and proved, after three years and 46,000 miles, that one man could sail around the world alone. This title is available in many other editions, including audio books.

Sea Soup: Phytoplankton
by Mary M. Cerullo; photos by Bill Curtsinger
Tilbury House, Gardiner, ME (1999)

Squids (Nature's Children)
by James Kinchen
Grolier Educational, Danbury, CT (1999)

Describes the physical characteristics, natural habitat, behavior, diet, and world distribution of squids.

Tentacles
The Amazing World of Octopus, Squid, and Their Relatives
by James Martin, Crown, New York, NY (1993; *out of print*)

Under the Sea
by Lindsay Knight; consulting ed. Dr. Frank H. Talbot
Nature Company Discoveries Library, Time-Life Books, New York, NY (1995)

Books for Teachers

Cephalopod Behaviour
by Roger T. Hanlon and John B. Messenger
Cambridge University Press, New York, NY (1998)

Excellent illustrations and straightforward language make this attractive book a useful addition to any natural history collection. The book is as readable for lay persons as for researchers.

Against the Tide
The Fate of the New England Fisherman
by Richard Adams Carey
Houghton Mifflin, Boston, MA (1999)

The Plundered Seas
Can the World's Fish Be Saved?
by Michael Berrill
Sierra Club Books, San Francisco, CA (1997)

The Restless Sea
by Robert Kunzig
W.W. Norton, New York (1999)

A spellbinding look at what's known about the world's oceans, and how that knowledge was assembled over the centuries. Chapter 8, with its look at the decline of Atlantic cod and the dangers of overfishing, is especially relevant to *Only One Ocean*.

The Search for the Giant Squid
by Richard Ellis
The Lyons Press, New York, NY (1998)

A wonderful montage of stories (real and legendary), references, illustrations, and museum displays about this elusive giant mollusk.

Marine Biology
by Peter Castro and Michael E. Huber
McGraw-Hill Higher Education, New York, NY (2000)

Intertidal Invertebrates of California
by R.H. Morris, D.P. Abbott, and E.C. Haderlie
Stanford University Press, Stanford, CA (1980)

A huge, comprehensive, and classic reference work on hundreds of invertebrates off California. Good for the school library.

Oceanography
An Invitation to Marine Science
by Tom Garrison
Wadsworth Publishing, Stamford, CT (1993)

Fish, Markets, and Fishermen
The Economics of Overfishing
by Suzanne Iudicello, Michael Weber, and Robert Wieland
Island Press, Washington, D.C. (1999)

Videos

Incredible Suckers
60 min.; Educational Broadcasting Corporation (2000)

A mesmerizing introduction to squids and their cephalopod relatives. Part of the PBS series *Nature*, with host George Page. See also www.pbs.org/wnet/nature/suckers/index.html.

Footsteps in the Sea
Growing Up in the Fisheries Crisis
21 min.; Bullfrog Films (1998)

Produced by the National Audubon Society's Living Ocean Program and the University of Tennessee, this film describes the challenges faced by many fishing communities embroiled in battles over diminishing resources.

Airing in February, 2002:
Empty Oceans, Empty Nets
Series of three, 60-min. programs; Public Broadcasting Service (2002)

An extremely promising documentary series on the state of global fisheries and efforts to restore them. Three discrete episodes cover the fisheries crisis, marine aquaculture, and innovative efforts at restoring fish populations and habitat. Let us know what you think, if you catch it!

TEDs: Turtle Excluder Devices
Safeguarding Shrimp Fisheries and Endangered Sea Turtles
16 min.; Earth Island Institute (1993)

Clear, interesting demonstration of this turtle-saving device in action.

Ancient Sea Turtles
Stranded in a Modern World
28 min.; Bullfrog Films/Earth Island Institute (1998)

The sea turtle's odyssey from hatchling to adult, and the perils along the way. Good introduction to TEDs. Package includes a useful teacher's insert.

Steller Sea Lions in Jeopardy
27 min.; Alaska Department of Fish and Game, et al. (1998)

Good overview of the threat to this important marine mammal from pollock overfishing and habitat destruction.

Where Have All the Dolphins Gone?
60 min.; The Video Project (1990)

The Marine Mammal Fund and the ASPCA document in grisly detail the deaths of dolphins in the course of commercial tuna fishing. **Not suitable for young students,** but a graphic illustration of the problem.

Secrets of the Ocean Realm
Set of five, 60-min. tapes; PBS Home Video (1998)

These five volumes from the Public Broadcasting Service explore the behavior of a variety of creatures in their undersea world. With its footage of squids, Volume 2, "Venom/Creatures of Darkness," is particularly useful for *Only One Ocean.*

Understanding Oceans
51 min.; Discovery Channel (1997)

Most of the water on Earth is gathered in a large, salty ocean that's pushed by the wind, pulled by the moon, and swirled as the planet spins. Explore this hidden world and meet some of its endangered creatures.

Internet Sites

Congress at Your Fingertips
www.congress.org/congressorg/dbq/officials

Find and communicate with your representatives in Congress and the Senate, and discover the committees and subcommittees on which they serve—look for "Fisheries"!

The Bridge: A Web-Based Resource for Ocean Sciences Educators
www.vims.edu/bridge/

A clearinghouse of the best K–12 ocean sciences education sites available online.

Cousteau Society
www.cousteausociety.org

The organization through which many millions of people have discovered and come to appreciate the fragility of life on our Water Planet. (See also the Dolphin Log in the Classroom section for information on that program.)

InciteScience!
www.incitescience.com

This science education site includes a list of marine science literature and resources.

MARE
www.lhs.berkeley.edu:80/MARE/

The Marine Activities, Resources & Education (MARE) program at the Lawrence Hall of Science is a dynamic, multicultural K–8 science program that transforms entire elementary and middle schools into laboratories for the exploration of the ocean.

Seafood Guides
www.montereyaquarium.org/efc/efc_oc/dngr_food_watch.asp
and
www.audubon.org/campaign/lo/

A growing number of seafood chefs are removing such popular items as Chilean sea bass or Atlantic swordfish from their menus, concerned about the dwindling populations of fish in the world's ocean. These two sites help consumers determine the most "ethical" fish to eat—or steer away from, as the case may be.

CephBase: Database on Cephalopods
www.cephbase.utmb.edu/index.html

This expansive site includes great reference material on squids and their kin. See the Image Gallery for *Loligo opalescens!*

Smithsonian's "Ocean Planet" Online Exhibit
www.seawifs.gsfc.nasa.gov/ocean_planet.html

A full exploration of the marine environment, with vital information about perils facing the ocean.

Understanding Our Oceans and Climate from Space
TOPEX/Poseidon
topex-www.jpl.nasa.gov/

A partnership between the U.S. and France to monitor global ocean circulation, study the tie between ocean and atmosphere, and improve global climate predictions.

"Oceans & Fisheries Update"
Talk of the Nation
National Public Radio Online
http://search.npr.org/cf/cmn/cmnpd01fm.cfm?PrgDate=04/20/2001&PrgID=5

With RealAudio software you can hear this hour-long program directly off the Web. Or you can order transcripts of the show, which aired April 20, 2001.

Seafood traders and news about the wholesale seafood market
www.seafood.com

National Marine Fisheries Service
www.nmfs.noaa.gov/

"U.S. Fish Harvesters Up on Financial Rocks"
Environmental News Service article
ens.lycos.com/ens/apr2000/2000L-04-25-01.html

Posters

Marine Mammals of the Gulf of the Farallones
Gulf of the Farallones National Marine Sanctuary
GGNRA, Fort Mason
San Francisco, CA 94123
(415) 556-3509

Mamiferos Marinos de Mexico
Pieter Folkens
940 Adams Street, Ste. F
Benicia, CA 94510

Shellfish/Edible Mollusks
Celestial Arts
P.O. Box 7123
Berkeley, CA 94707
(510) 559-1600

Olympic Coast National Marine Sanctuary Dedication
Olympic Coast National Marine Sanctuary
138 W. First Street
Fort Angelos, WA 98362
(206) 457-6622

Save Our Seas Curriculum Poster
California Coastal Commission
45 Fremont Street, Ste. 2000
San Francisco, CA 94105
(415) 904-5206

Many marine sanctuaries and estuarine reserves provide educational posters for teachers. Contact any of the following for more information. These estuaries and marine sanctuaries can also be a great source for local information.

National Estuary Program Contacts

1
Puget Sound, WA
Puget Sound Water Quality Authority
(206) 407-7300

2
Tillamook Bay, OR
Tillamook Bay National Estuary Program
(503) 842-9922

For information on estuaries 1 and 2, you may also contact:
U.S. EPA, Seattle, WA
Surface Water Branch
(206) 553-4183

3
San Francisco Estuary, CA
San Francisco Estuary Project
San Francisco Bay Regional Water Quality Control Board
(510) 286-0625

4
Santa Monica Bay, CA
Santa Monica Bay Restoration Project
(213) 266-7515

For information on estuaries 3 and 4, you may also contact:
U.S. EPA, San Francisco, CA
Watershed Protection Branch
(415) 744-1953

5
Corpus Christi Bay, TX
Corpus Christi Bay National Estuary Program
(512) 985-6767

6
Galveston Bay, TX
Galveston Bay National Estuary Program
(713) 332-9937

7
Barataria-Terrebonne Estuarine Complex, LA
Barataria-Terrebonne National Estuary Program
(504) 447-0868
(800) 259-0869

For information on estuaries 5–7, you may also contact:
U.S. EPA, Dallas, TX
Water Quality Branch
(214) 655-7135

8
Tampa Bay, FL
Tampa Bay Estuary
National Estuary Program
(813) 893-2765

9
Sarasota Bay, FL
Sarasota Bay National Estuary Program
(813) 361-6133

10
Indian River Lagoon, FL
Indian River Lagoon National Estuary Program
(407) 984-4950

11
Albemarie-Pamilco Sounds, NC
Albemarie-Pamlico Estuarine Study
NC Department of Environment, Health, and Natural Resources
(919) 733-0314

For information on estuaries 8–11, you may also contact:
U.S. EPA, Atlanta, GA
Wetlands, Oceans, and Watershed Branch
(404) 347-1740

12
Delaware Inland Bays, DE
Delaware Inland Bays Estuary Program
Delaware Department of Natural Resources and Environmental Control
(302) 739-4590

13
Delaware Estuary, DE, PA, and NJ
Delaware Estuary Program
U.S. EPA, Philadelphia, PA
(215) 597-9977

For information on estuaries 12 and 13, you may also contact:
U.S. EPA, Philadelphia, PA
Environmental Assessment Branch
(215) 597-1181

14
New York-New Jersey Harbor, NY and NJ
New York Department of Environmental Conservation, Albany, NY
(518) 485-7786
New Jersey Department of Environmental Protection and Energy
(609) 292-1895

15
Long Island Sound, NY and CT
Long Island Sound Office
(203) 977-1541

16
Peconic Bay, NY
Peconic Bay Program
Suffolk County Department of Health Services
(516) 852-2080

For information on estuaries 13–16, you may also contact:
U.S. EPA, New York, NY
Marine & Wetlands Protection Branch
(212) 264-5170

17
Narragansett Bay, RI
Narragansett Bay Project
Rhode Island Department of Environmental Management
(401) 277-3165

18
Buzzards Bay, MA
Buzzards Bay Project
(508) 748-3600

19
Massachusetts Bays, MA
Massachusetts Bays Program
(617) 727-9530

20
Casco Bay, ME
Casco Bay Estuary Project
(207) 828-1043

For information on estuaries 15 and 17–20 , you may also contact:
U.S. EPA, Boston, MA
Water Quality Branch
(617) 565-3531

21
San Juan Bay, PR
PR Environmental Quality Board
(809) 751-5548
Puerto Rico Department of Natural Resources and Environment
(809) 724-5516

For information on estuary 21, you may also contact:
U.S. EPA, Caribbean Field Office
Santorce, PR
(809) 729-6921
U.S. EPA, New York, NY
Marine & Wetlands Protection Branch
(212) 264-5170

National Marine Sanctuaries

Channel Islands
National Marine Sanctuary
113 Harbor Way
Santa Barbara, CA 93109
(805) 966-7107 fax (805) 568-1582

Cordell Bank
National Marine Sanctuary
Fort Mason, Building 201
San Francisco, CA 94123
(415) 556-3509 fax (415) 556-1419

Fagatele Bay
National Marine Sanctuary
P.O. Box 4318
Pago Pago, American Samoa 96799
(684) 633-5155 fax (684) 633-7355

Florida Keys
National Marine Sanctuary
9499 Overseas Highway
Marathon, FL 33050
(305) 743-2437 fax (305) 743-2357

*Key Largo
National Marine Sanctuary
P.O. Box 1083
Key Largo, FL 33037
(305) 451-1644 fax (305) 451-3193

*Looe Key
National Marine Sanctuary
Rt. 1, Box 782
Big Pine Key, FL 33043
(305) 872-4039 fax (305) 872-3860

**Part of Florida Keys National Marine Sanctuary*

Flower Garden Banks
National Marine Sanctuary
1716 Briarcrest Drive, Suite 702
Bryant, TX 77802
(409) 847-9296 fax (409) 845-7525

Gray's Reef
National Marine Sanctuary
P.O. Box 13687
Savannah, GA 31416
(912) 598-2345 fax (912) 598-2367

Gulf of the Farallones
National Marine Sanctuary
Fort Mason, Building 201
San Francisco, CA 94123
(415) 556-3509 fax (415) 556-1419

Hawaiian Islands Humpback Whale
National Marine Sanctuary
1305 East-West Highway
SSMC4, 12th Floor
Silver Springs, MD 20910
(301) 713-3141

Monterey Bay
National Marine Sanctuary
299 Foam Street, Suite D
Monterey, CA 93940
(408) 647-4201 fax (408) 647-4250

Stellwagen Bank
National Marine Sanctuary
14 Union Street
Plymouth, MA 02360
(617) 982-8942

Monitor (sanctuary named after a sunken ship)
National Marine Sanctuary
NOAA
Building 1519
Fort Eustis, VA 23604-5544
(804) 878-2973 fax (804) 878-4619

Proposed Sanctuaries:
Sanctuaries and Reserves Division
National Oceanic and Atmospheric Administration (NOAA)
1305 East-West Highway
SSMC4, 12th Floor
Silver Springs, MD 20910
(301) 713-3125

Assessment Suggestions

Selected Student Outcomes

1. Students are able to describe the relative percentages of the Earth that are covered by land and ocean, and demonstrate increased understanding of the importance of the ocean in making the planet habitable, including providing oxygen and food. Students are able to communicate a variety of ways in which people depend on the ocean for food, oxygen, recreation, or their livelihoods.

2. Students demonstrate awareness that life and resources in the ocean are limited and unequally distributed; most are concentrated in a relatively small portion of the ocean. Students deepen their understanding and appreciation of the importance and fragility of natural resources.

3. Students show increased understanding of invertebrate anatomy; especially the idea that in all animals, form (structure) is closely related to function. By studying one animal—squid—in depth, students gain insight into adaptations needed to survive in the open ocean, and the scientific importance of studying organisms together with the environments in which they live.

4. Students are better able to communicate the complexity and difficulty of many environmental problems. Solutions they propose should reflect complex thinking that takes into account "trade-offs," real-world constraints, and other obstacles.

5. Students demonstrate an increased ability to understand, visualize, chart, and explain both data and their own reasoning mathematically, through originating and interpreting graphs as well as by working with and analyzing large numbers, proportions, fractions, percentages, and decimals.

Built-In Assessment Activities

Prior Knowledge. In Activity 1, Session 1, students brainstorm in small groups what they already know about the ocean. The teacher can circulate, making notes about the levels of prior knowledge students bring to the unit (as well as information about social skills, language ability, and ability to work in a group). When groups share ideas to contribute to class charts, these can provide a record of the general level of their prior knowledge. The teacher can gauge the level of instruction accordingly—and can bring this chart out at the end of the unit as a dramatic way of showing students (and the teacher herself!) how much everyone has learned. (Addresses Outcomes 1, 2)

Circle Graphs. In Activity 1, Session 2, students create circle graphs representing their understanding of several concepts related to the Earth's productivity. The teacher can collect these finished products and use them to assess students' understanding of these concepts. Teachers can also collect students' lists of five things they learned from "Think, Pair, Share," and compare these notes to the circle graphs to see if they support one another. (Outcomes 2, 5)

Mini-Books. In Activity 1, Session 3, students create "Mini-Books" summarizing what they've learned. These are ideal portfolio items that can be used as writing samples and to assess conceptual understanding. The books can be compared with the circle graphs to see if students are able to explain similar concepts consistently in different formats/genres. If students create drawings, posters, brochures, or booklets, these can also be used for assessment purposes. (Outcomes 1, 2)

Anticipating Squids. In Activity 2, Session 1, group Anticipatory Charts about squids record students' initial knowledge and curiosity. Since they share prior knowledge first with a partner, then in a foursome, then with the class, students' charts can reveal their ability to acquire information from peers, integrate it into their own knowledge base, and revise their thinking. Charts can be collected and used as part of pre/post assessment. Matching students' initial questions with later products can be a valuable window into their thinking. (Outcome 3)

Anatomical Charts. In Activity 2, Session 2, students create and label internal and external squid anatomy diagrams. These can become part of their portfolios. The diagrams can be scored in two ways: a) in a formal way, as to whether all the parts of the squid are located and labeled appropriately, or b) informally (but at least as importantly) by providing valuable insights into students' ability to record important information and to translate their observations into drawings and text to help them recall the information. (Careful note-taking is an essential real-world science skill, and sketches and labeled drawings are key elements.) (Outcome 3)

Some students who have difficulty demonstrating what they've learned via more traditional assessments may excel when asked to organize their thoughts graphically. If this is true of some students, the teacher should take special note and be sure to point it out to them in a positive way!

Debriefing. The Pictionary, Jeopardy, and other debriefing activities that make up the third session of Activity 2 all provide opportunities to note what individual students have learned and understood. The teacher can keep a running notebook of each student's responses. These types of engaging class activities are particularly useful for eliciting and identifying remaining misconceptions that may need to be addressed at some point in the unit. (Outcome 3)

Squid Fishery Presentations. In the last session of Activity 2, student presentations about the squid fishery can provide insights into their comprehension of information on the profile cards and, more importantly, into their critical-thinking skills. The teacher can take note when students present information or points of view (especially potential solutions) that don't just come directly from the cards, but from their own ideas after grappling with complex issues. The quality of their presentations is also an indication of their ability to take part in scientific and social discourse. (Outcome 3)

Fishery Projects. The complex projects students work on in groups and individually in Activity 3 provide a fertile field for assessment. Students produce many draft and final products, large and small, and such multiple indicators of students' progress are especially helpful in making a full assessment of their progress. The posters students create are indicators of their thinking and comprehension. Predicting, for example, is a complex critical-thinking skill, and student predictions about the future of fisheries should be supported with data and logical reasoning. Student posters, and teacher notes on student presentations, can also indicate if students did further research—going beyond the information cards. Students' reactions and responses to questions are helpful in assessing their grasp of the information and the development of their own ideas. And the questions students ask can be powerful indicators of their depth of understanding of complex issues. (Outcomes 1, 2, 4, 5)

Fish Math. Embedded in and utilized throughout the ocean fisheries activity are a number of analytic mathematics tools. Insight into math abilities and comprehension can be gained by considering students' use and interpretation of graphs and other representational methods—and especially by their own manipulations and explanations of data, including their creation of graphs and/or other ways to depict and analyze information. Additional insight into students' math literacy can spring from their work with large numbers, their ability to use and apply reasoning and notation relating to scale and scale-related legends, as well as their facility with percentages, ratios and proportions, fractions, and/or decimals. (Outcome 5)

Recommending Solutions. In Activity 3, Session 4, the recommendations that emerge from the fisheries conference are indicators of the level of students' thinking. The most sophisticated solutions may be those that students acknowledge are imperfect or that include important trade-offs and/or compromises. Their recommendations may also reveal how well students understand and apply the concept of "not wasting" (as modeled in the squid activity) to their fishery-management ideas. In the final "Think, Pair, Share: My Values and Behavior," the sheets individual students write on about their choices can be collected. They can be used as an assessment, not of right or wrong answers, but of how students support their decisions and whether they express their ideas with conviction. (Outcomes 2, 4)

Additional Assessment Ideas

I Used to Think. If students are asked to do "Wrap Up #2" in Activity 2, Session 2, or "I Used to Think but Now I Know" Books in Session 3, their writing about a particular adaptation can be used as a specific assessment that samples their depth of knowledge in one area of their choice. It's important to provide opportunities for students to demonstrate that they understand both the overview/big picture of the concepts *and* that they have attained some depth of understanding—or even mastery—of some subset of specific concepts. Learning both "facts" and how the facts fit together is important. (Outcome 3)

The Ocean of Creativity. After completing this unit, consider having students, either as individuals or in groups, come up with a creative effort in consultation with you, the teacher. Their goal is to portray the new understandings they've gained about the ocean. This could be a poem, a rap, a song, a drawing, a large paper mural, a short drama or role-play, even a dance of the squid—so long as it gets across one or more of the main lessons of the unit. (Outcomes 1, 2, 3, 4)

Many additional assessment tools can be drawn from the "Going Further" ideas provided after each activity.

Literature Connections

The fiction we've selected as literature connections for *Only One Ocean* range from ocean-specific stories to broader ecological tales. *The Great Kapok Tree*, for example, carries a strong conservation message. *The Magic School Bus on the Ocean Floor* gives students a look at the entire ocean ecosystem.

In the listings below, the grade-level estimates reflect both interest level and reading level. Many of the books can be enjoyed by a wide age range—older students can read them on their own and they can be read aloud to younger students. Some books (like the rhyming "alphabet of the ocean," *Into the A, B, Sea*) are marketed for a young audience but can be inspired tools for helping middle school students organize their thoughts, figure out what's important, do research, and use words in innovative ways. They're also excellent for encouraging language development in English-language learners.

Please be sure to see the many excellent resource and reference books for students and teachers listed in the "Resources" section on page 163. You may also want to refer to the GEMS literature connections handbook, *Once Upon a GEMS Guide: Connecting Young People's Literature to Great Explorations in Math and Science.* It lists books according to science themes and mathematics strands, as well as by connections to GEMS guides. We're always looking for titles to add to future editions of *Only One Ocean* and *Once Upon A GEMS Guide.* Please let us know how these suggestions worked for you, and send us your nominations for more books about the interconnectedness of our one, vast ocean.

The Case of the Missing Cutthroats
An Eco Mystery
by Jean Craighead George
HarperCollins, New York, NY (1996)
(originally published by E.P. Dutton in 1975 as *Hook a Fish, Catch a Mountain*)
Grades: 3–7

After Spinner Shafter catches a cutthroat trout in the Snake River, she and her cousin Alligator search the nearby mountains to determine where the endangered fish came from and how it survived. The author has written several other worthwhile ecological mysteries, including *Who Really Killed Cock Robin?* and *The Missing 'Gator of Gumbo Limbo.*

20,000 Leagues Under the Sea,
by Jules Verne
1870: many contemporary editions available
Grades 3–Adult

The classic and still-compelling yarn. Plenty of creative oceanography and exaggeration... but great fun and surprisingly relevant.

The Great Kapok Tree
A Tale of the Amazon Rain Forest
by Lynne Cherry
Harcourt Brace Jovanovich, San Diego (1990)
Grades: K–4+

The many animals that live in a great kapok tree in the Brazilian rainforest entreat an ax-wielding man to understand the importance of their tree home, and persuade him not to cut it down. Although this picture book is meant for a younger audience, it's useful to demonstrate that in the same way one tree in a forest is important to many other life forms, one species in the ocean—the squid, for instance—is important to many organisms.

Get Connected–Free!

Get the *GEMS Network News*,

our free educational newsletter filled with...

- **updates** on GEMS activities and publications
- **suggestions** from GEMS enthusiasts around the country
- **strategies** to help you and your students succeed
- **information** about workshops and leadership training
- **announcements** of new publications and resources

Be part of a growing national network of people who are committed to activity-based math and science education. Stay connected with the* GEMS Network News. *If you don't already receive the* Network News, *simply return the attached postage-paid card.

For more information about GEMS call (510) 642-7771, or write to us at GEMS, Lawrence Hall of Science, University of California, Berkeley, CA 94720-5200, or gems@uclink4.berkeley.edu.

Please visit our web site at www.lhsgems.org.

GEMS activities are effective and easy to use. They engage students in cooperative, hands-on, minds-on math and science explorations, while introducing key principles and concepts.

More than 70 GEMS Teacher's Guides and Handbooks have been developed at the Lawrence Hall of Science – the public science center at the University of California at Berkeley – and tested in thousands of classrooms nationwide. There are many more to come – along with local GEMS Workshops and GEMS Centers and Network Sites springing up across the nation to provide support, training, and resources for you and your colleagues!

Yes!

Sign me up for a free subscription to the

GEMS Network News

filled with ideas, information, and strategies that lead to Great Explorations in Math and Science!

Name____

Address______________________________________

City______________________ State________ Zip____________

How did you find out about GEMS? (Check all that apply.)

❐ word of mouth ❐ conference ❐ ad ❐ workshop ❐ other: ____________________

❐ In addition to the *GEMS Network News*, please send me a free catalog of GEMS materials.

GEMS
Lawrence Hall of Science
University of California
Berkeley, CA 94720-5200
(510) 642-7771

Ideas
Suggestions
Resources

Sign up now for a free subscription to the *GEMS Network News!*

that lead to Great Explorations in Math and Science!

101 LAWRENCE HALL OF SCIENCE # 5200

1-61571-25775-62-X

NO POSTAGE NECESSARY IF MAILED IN THE UNITED STATES

BUSINESS REPLY MAIL
FIRST-CLASS MAIL PERMIT NO 7 BERKELEY CA

POSTAGE WILL BE PAID BY ADDRESSEE

UNIVERSITY OF CALIFORNIA BERKELEY
GEMS
LAWRENCE HALL OF SCIENCE
PO BOX 16000
BERKELEY CA 94701-9700

Into the A, B, Sea
An Ocean Alphabet
by Deborah Lee Rose; illustrated by Steve Jenkins
Scholastic, New York, NY (2000)
Grades: K–3+

Delightful rhyming text combines with vivid cut-paper illustrations to give readers a tour of the ocean and its inhabitants—from Anemones to Zooplankton. Each verse ("...where kelp forests sway and leopard sharks prey....") succinctly captures its creature's unique attributes. A glossary provides further information on each animal, and the book includes a teacher's supplement.

Island of the Blue Dolphins
by Scott O'Dell; illustrated by Ted Lewin
Houghton Mifflin, Boston (1990)
Grades: 5–12

Based on a true story, this is the tale of a young Native American girl left alone on a beautiful but isolated island off the coast of California for 18 years. Interweaving descriptions of the island, fish and ocean vegetation, and animals and plants, it tells not only of her enormous courage and self-reliance in learning to survive, but also of how she found a measure of happiness in her solitary life.

Just A Dream
by Chris Van Allsburg
Houghton Mifflin, Boston (1990)
Grades: All Ages

After dreaming about a future Earth devastated by pollution, Walter begins to understand the importance of caring for the environment. Unique and evocative pictures of what our future may hold provide the powerful backdrop as young Walter becomes enlightened and changes his thinking and actions.

The Magic School Bus—On the Ocean Floor
by Joanna Cole; illustrated by Bruce Degen
Scholastic, New York (1992)
Grades: 1–6

Her students may be expecting a trip to the beach, but in her own predictable style Ms. Frizzle takes her class on a field trip under the ocean. The class explores many different ocean habitats and learns about the organisms in each. In one report along the book's margins, a student discusses how all the oceans of the world are connected to form "one world ocean."

The Old Ladies Who Liked Cats
by Carol Greene; illustrated by Loretta Krupinski
HarperCollins, New York (1991)
Grades: K–6

When a group of old ladies are no longer allowed to let their cats out at night, the delicate balance of their island ecology is disturbed—with disastrous results. Based on Charles Darwin's story about clover and cats, this ecological folk tale demonstrates the interrelationships of plants and animals. (Note: Some teachers may wish to balance this tale with a discussion about the impact of outdoor domestic cats on birds and other wildlife.)

Out of the Ocean
by Debra Frasier
Harcourt Brace and Company, San Diego (1998)
Grades: Preschool–5+

As a young girl and her mother walk along an eastern Florida beach, they marvel at the many treasures cast up by the sea and the wonders of the world around them. Detailed and illustrated pages at the end of the book give information about the items found. Ocean currents are discussed when one of the items proves to be a bottle with a note inside. Although intended for younger students, the book extols the riches of the ocean for everyone.

Shark Beneath the Reef
by Jean Craighead George
HarperCollins, New York (1989)
Grades: 5–12

Tomás, 14 years old, has two loves: fishing and school. He's supported in each of these passions by his proud fisherman grandfather and his caring high school science teacher. Tomás comes from a family of shark fishers, on the island of Coronado on the Sea of Cortez, whose livelihood is threatened by Japanese factory fishing boats and government plans for tourism. The oceanic environment flows through the book as Tomás observes the activity in a tide pool or tracks a fish underwater, giving a real sense of the interrelation of marine life and its habitats.

Waterman's Boy
by Susan Sharpe
Bradbury Press, New York (1990)
Grades: 3–6

Two boys from a small town on Chesapeake Bay defy their fisher parents to help a scientist interested in cleaning up the water for the benefit of animals, plants, and people. The book explores how some people who rely on the ocean for a living fear outsiders who appear to be interfering with a way of life.

Summary Outlines

Activity 1: Apples and Oceans

Getting Ready

1. Make two blank Brainstorm Charts (one each for Sessions 1 and 2) and (optional) one blank Question Chart.
2. Make and hang "Land" and "Ocean" posters or create transparencies.
3. Post discussion questions from page 18.
4. Write out Key Concept for this activity on chart paper:
   - **Most of our planet is covered by ocean, but only a small fraction of the ocean supports large concentrations of life.**
5. Set up inflatable globe and (optional) ocean posters.
6. Gather (optional) ocean-sounds audiocassette, cassette player, and ocean photos for brainstorm.
7. Read through all sessions, especially Session 2: Apples and Oceans.
8. Buy and wash the apples.

Session 1: Brainstorm—Our "Planet Ocean"

1. Arrange pairs of students in groups of four or six, pass out pictures and photos, and play ocean-sounds cassette. Have student groups brainstorm posted questions and record their own.
2. Discuss the questions as a class, recording on Brainstorm Chart for Session 1 (and optional Question Chart).
3. Show the traditional "map" and "ocean" views on the globe; discuss the important concepts on page 19.

Session 2: Apples and Oceans

1. Explain how an apple will serve as a model for the planet and its limited resources.
2. Have students clear their tables and sit in pairs. Give each pair an apple, a knife, a paper plate, a paper towel, and a number of colored markers. Keep a set for yourself.
3. Introduce "Land" and "Ocean" posters and recap the modeling concept.
4. Assign or have pairs choose a "land" person and an "ocean" person. Identify land and ocean people and briefly model what each will do.
5. Model, have students duplicate, and then explain the steps for each of these LAND slices of the apple/planet and counterpart circle-graph sections (pages 21–25):
   - Slice One: $\frac{1}{4}$ of the Planet is LAND;
   - Slice Two: $\frac{1}{8}$ of the Planet is Uninhabitable Land;
   - Slice Three: $\frac{1}{8}$ of the Planet is Habitable Land;
   - Slice Four: $\frac{1}{32}$ of the Planet is Farmable Land;
   - Slice Five: $\frac{3}{10,000}$ of the Planet is Land with Drinkable Water.

6. Pause to introduce the OCEAN portion of the activity and let land and ocean people switch roles. Explain that three-fourths of the planet is OCEAN. Model, have students duplicate, and explain the steps for each of these two ocean slices and their counterpart circle-graph sections (pages 26–28):

- Slice Six: $\frac{1}{16}$ of the Planet is Productive Coastal Zone;
- Slice Seven: $\frac{3}{4,000}$ of the Planet is Upwelling Zone.

Present in-depth session on ***upwelling*** *and/or introduce the* ***photic zone,*** *if you choose.*

Wrap-Up

1. Have students hold up the slivers representing the planet's **drinkable water** and **upwelling zones**, and discuss what they've learned about these limited resources.

2. Have students clean up their materials and recycle the apples.

Think, Pair, Share: What Did We Learn?

1. Prepare students to review some of the details they've learned. Have them **"think"** of what most surprised or interested them about "Planet Ocean" and write down at least three things.

2. Have students **"pair"** with a partner and **"share"** their notes.

3. Have each pair list five new things they learned about the land, the ocean, or the planet as a whole.

4. Initiate class brainstorm for sharing the lists; record students' thoughts on Brainstorm Chart for Session 2. As you write, groups add items from other groups' lists to their own. Compare new chart with original class Brainstorm Chart from Session 1.

5. Hold up Key Concept for this activity and have one or more students read it aloud. Briefly discuss its relevance to today's activities. Post concept for students to revisit.

Session 3: Creative Writing

A. Mini-Books on the Ocean

1. Lead students through the construction of individual mini-books.

2. Describe how to title, illustrate, and write in the books to reflect their newfound knowledge about the ocean, the land, and the planet.

3. Allow students time to design and fill their books and share them with others.

B. Posters, Journal Writing, Comic Books, and Travel Brochures

1. Brainstorm possible topics and scenarios from this activity for students to use in creative-writing assignments.

2. Have students share their work through poster sessions or informally.

Activity 2: Squids—Outside and Inside

Getting Ready

1. Purchase squids for dissection. Keep frozen until use.

2. Start class Squid Statements Anticipatory Chart, using student master as example.

3. Copy the Squid Statements Anticipatory Chart master for each small group.

4. Start class About Squids Chart.

5. Draw squid chart #1, External Squid, and squid chart #2, Internal Squid. (Decide if you'll faintly trace internal organs on Internal Squid diagram and laminate.) Draw squid chart #3, What We Learned.

6. Decide if you'll play Squid Jeopardy, and if so, copy unlabeled Internal and External Squid Diagram student sheets for each student.

7. Make class Squid Interest-Group Chart.

8. Copy Squid Interest-Group Profiles master and cut into individual profiles.

9. Write out Key Concepts for this activity on chart paper:

- **Pelagic creatures are organisms living in the open ocean.**
- **Looking closely at an animal like the squid can tell us a lot about the adaptations needed to survive and thrive as a pelagic creature.**
- **Many people depend on squids for food or for their livelihood. More discussion among these people will help create solutions to the problem of diminishing squid populations.**

10. Read through and practice a squid dissection for yourself, using (optional) Squid Dissection Summary Outline for Teacher's Notes.

11. Determine when you'll cook the squids and arrange for help if needed.

12. Decide if and how you'll invite others to participate.

13. Prepare plates for each pair of students, with scissors, hand lens, ruler, toothpicks, and (just before activity) squids.

14. Cue squid-footage videotape.

15. Have students thoroughly wash their hands.

Session 1: What about Squids? Anticipatory Charts

Squid Statements

1. Review the definition and significance of upwelling.

2. Explain that the class will be studying (up close!) an important ocean organism—the squid.

3. Divide students into small groups and distribute one Squid Statements Anticipatory Chart to each group.

4. Have each group elect a recorder, and start the groups discussing the statements on the chart.

5. After charts are filled in, play squid video footage without sound; have students

quietly discuss the statements again in light of what they see. After the video, allow students to add to their Anticipatory Charts.

6. Discuss all ideas and record on classroom Squid Statements Anticipatory Chart.

Think, Pair, Share: "About Squids"

1. Lead the class in a Think, Pair, Share activity:

   a. Have each student create an About Squids chart, **"think"** about what she knows and would like to find out, and record her ideas.

   b. Have students **"pair"** up to compare and discuss their ideas.

   c. Have each pair join another pair to **"share"** their combined answers.

2. Lead a class discussion and record all ideas on the class About Squids Chart.

Session 2: Hands-On Squid

After you've read Special Notes for the Teacher and Prelude to the Dissection, discuss the upcoming dissection with the class. Emphasize respect, taking instruction, and working in pairs.

The Dissection

1. Post squid chart #1, External Squid, and squid chart #2, Internal Squid; you'll be drawing in the anatomy as each part is introduced in the dissection.

2. Pair students up and distribute plates assembled with squids and tools. Have paper towels ready. Pass out one sheet of blank paper per pair, for students to label "External Squid."

External Anatomy: Arms, Head, Mouth, Eyes, and More

1. Be sure all students have washed their hands and that everyone's squid (including yours) is correctly oriented on the plate—paler (ventral) side up.

2. Ask students to look at the whole squid, noting connection of the "head" to the body.

3. Describe, have students locate, and then discuss each part of the squid's external anatomy covered here (pages 55–57): arms, tentacles, suckers, beak, buccal mass, radula, esophagus, salivary glands, and eye lens. As you draw each part on class chart, referring to your Squid Dissection Summary Outline, have students draw on their diagrams.

External Anatomy: Mantle, Color, Fins, and More

1. Describe, have students locate, and then discuss the parts and adaptations covered here (pages 58–60): mantle, chromatophores, countershade coloration, fins, and funnel. Again, as you draw each part on class chart, have students draw on their diagrams.

2. Ask students to measure their squids.

3. Have students peek inside the mantle and guess what they'll find next, during the internal anatomy dissection. Record the vocabulary mentioned in their ideas.

Internal Anatomy

1. Be sure everyone's squid is correctly oriented on the plate: paler ventral side up.

2. Pass out blank paper for each pair to label "Internal Squid."

3. As you begin the internal dissection, alert students to cut carefully through just the mantle and not the underlying organs. Have them spread open the mantle and compare their squids with others in the class.

4. Briefly ask students what differences they've observed; they'll likely include "male" and "female." Explain principal differences between male and female squids as described on pages 62–63.

5. Have students try to identify the sex of their squids (this can be tricky; see sidebar on page 63), and be sure everyone has the opportunity to see one of each. Tally number of male and female squids in the class.

6. Continue with reproductive anatomy that follows, OR skip to briefer description on page 65 and then move on to "Digestive System," below.

Females Only

Describe, have students locate, and then discuss each part of the female squid's reproductive anatomy covered here (pages 62–65): nidamental glands, ovary, oviduct, oviducal gland, eggs, and accessory nidamental gland. As you draw each part on class chart, referring to your Squid Dissection Summary Outline, have students draw on their diagrams.

Males and Mating

Describe, have students locate, and then discuss each part of the male reproductive parts covered here (pages 62–63 and page 65): vas deferens, spermatophoric gland, and terminal organ. Have students draw on their diagrams as you add to your chart.

Squid Reproduction Wrap-Up

1. If using video with squid reproduction footage, play that section again—without sound—and explain how squids "shoal" to mate and spawn (lay eggs).

2. Describe the mating process, referring to Note on Squid Reproduction on page 65.

Digestive System

Describe, have students locate, and then discuss each part covered here (pages 65–66): stomach, caecum, intestine, anus, funnel retractor muscles, and digestive gland. Have students draw on their diagrams as you draw on your chart.

Ink Sac

1. Have students remove the ink sac with care, so it doesn't puncture.

2. Explain the ink sac's purpose. Pierce one in water (don't use a student's) to demonstrate diffusion of the ink.

3. Have students carefully set aside their ink sacs for later use.

Respiratory and Circulatory Systems

1. Explain or review the concept that marine organisms draw oxygen from water.

2. Repeat the modeling and other steps for the parts covered here (page 67): gills, branchial hearts, and systemic heart. Add to your chart and have students add to their diagrams.

Support System

1. Describe the squid pen and its relationship to external shells in other mollusks.

2. Explain how to remove the pen from the squid.

3. Students can now dip their "pens" in the ink from their ink sacs and write their names on their plates or drawings.

The Nervous System

1. Describe, have students locate, and then discuss the parts covered here (page 68): brain and nerves.

2. To expand on the unique qualities of the squid's nerve fibers (axons), refer to sidebar on page 68 and to page 152 of "Behind the Scenes."

Wrap-Up

1. Collect students' illustrated internal and external anatomy diagrams, which become part of student portfolios used for assessment.

2. Store, discard, or prepare to cook (see below) the dissection squids.

3. If you choose, have students review and write about one of the squid's adaptations.

Calamari Festival *(Cooking instructions appear on pages 70–71.)*

1. Clean undissected squids and/or show students how to clean their dissected squids.

2. Have students discard all internal organs except the pen and eye lens, which they may keep.

3. Collect cleaned squids (to be cooked) and dissection scissors (to be washed).

4. Have students clean off their tables, wash their hands, and prepare to eat the calamari.

5. Hold up and post the two Key Concepts for this session **(Pelagic creatures are organisms living in the open ocean**.... and **Looking closely at an animal like the squid....)** and have students read aloud.

Session 3: Debriefing Activities

Squid Pictionary

1. Write squid parts on separate slips of paper and put in a box.

2. Have a student pick a slip and—**without naming the part**—draw it on squid chart #3, What We Learned.

3. Allow class to guess the name of the part and describe its function.

4. Have students write the part on their squid diagrams.

Squid Jeopardy

1. Distribute one unlabeled Internal and one unlabeled External Squid Diagram student sheet to each student.

2. Describe the *function* of one squid part, without naming it.

3. Call on a student to name the part **in the form of a question.**

Video Highlights and Reruns

1. Play squid video again, without sound.

2. Have students narrate each scene, pointing out body parts and describing squids' adaptations to their habitat.

Alternate version:

1. Have students list adaptations of squids in video.

2. Divide students into small groups.

3. Assign, or allow students to choose, one listed adaptation per group.

4. Cue video to each adaptation and have each group make a presentation to the class.

"I Used To Think but Now I Know" Books

1. Have students complete the last ("What we know to be true") column of their individual Anticipatory Charts.

2. Lead a class brainstorm and record students' responses.

3. Ask students to silently choose a favorite fact about squids and match it with something from the "What we think we know" column of the chart.

4. Have them write their facts in an "I used to think_______but now I know______" sentence and add an illustration.

Session 4: Squid Fishery Symposium

What's My Concern?

1. Have students reflect briefly on what they've learned about squids. Do they appreciate squids more than before? Why do people eat squids? Who catches and eats squids?

2. Ask what should be done about the demand for squids going up while the squid population goes down.

3. Prepare students to represent different squid fishery "interest groups" at Squid Fishery Symposium.

4. Divide class into six groups, each representing one of six "Squid Interest Groups" on the chart. Tape up the chart and show the names of the interest groups, **keeping questions covered**. Tell student groups they'll be given a description of their interest group's point of view and will answer questions from that viewpoint.

5. Assign, or have each group choose, each of the following roles:

a. **Reader**, to read "Squid Interest-Group" to group;

b. **Recorder**, to take notes about group's viewpoint as they discuss it; and

c. **Presenter**, to present the viewpoint at Squid Fishery Symposium.

6. Pass out a Squid Interest-Group Profile and one sheet of lined paper to each group. Discussion begins, with Reader and Recorder doing their parts.

Share with the Class

1. Have Presenters from each group tell the rest of the class about their group's point of view. They should paraphrase or act out the viewpoint with help from their group members.

2. Give the class a few minutes to ask questions of each presenting group.

3. Uncover the questions on the class chart and lead a discussion to fill in the blanks (in any order). Record answers on class chart.

4. Hold up and post Key Concept for this session **(Many people depend on squids....)** and have one or more students read aloud.

5. Prepare the class to learn, in the next activity, about a major worldwide environmental issue: overfishing (overharvesting).

Activity 3: What's the Catch?

Getting Ready

1. Make one copy of **each of the six different Fishery Information Cards:** What's Known about It?, What's for Dinner?, Where in the World?, How Are They Caught?, What's Happening with the Fishery?, and What's the Big Deal? **for each of the five different fisheries** (Swordfish, Tuna, Squid, Pollock, and Shrimp)**.** You'll have six cards for Swordfish, six for Tuna, six for Squid, six for Pollock, and six for Shrimp (a total of 32 sheets of paper).

2. Make **one copy of all six Fishery Assignments for each small group of Fishery Experts. Include the World Map Handout with Assignment #3, "Where in the World?";** decide whether to label the map yourself or have the students with this assignment do it. Paperclip assignments in sets of six for each group.

3. If possible, obtain five or so marine-life reference books (see "Resources").

4. Decide if you'll invite others to the World Fisheries Conference. The class can design a flyer advertising the event.

5. Read Overview of the World's Ocean Fisheries and Dolphins and Your Tuna-Fish Sandwich. Write key vocabulary on chart paper. Include ***sustainable, maximum sustainable yield (MSY), target species, bycatch (incidental take), habitat destruction, seafood, overharvesting/overfishing.***

6. Make five copies of Overview of the World's Ocean Fisheries on colored paper.

7. Make copies of Dolphins and Your Tuna-Fish Sandwich handout for all students.

8. Write out Key Concepts for this activity in large, bold letters on separate strips of chart paper:

- **Most large commercial ocean fisheries flourish where the interaction of currents and sunlight provide a productive environment.**
- **Most of the ocean fisheries in the world are severely threatened due to overfishing or habitat loss, and most commercial fishing results in significant "bycatch."**
- **Personal choices about what we eat can influence public policy and the sustainability of fisheries. Scientific information should be used to help make wise choices.**

9. Just before presenting Session 1, shop for seafood to serve at Seafood Smorgasbord.

10. Prepare trays for each table: unlabeled seafood samples, toothpicks, napkins, sheets of paper for students to draw guesses on, pencils, colored markers, and (optional) marine-life reference book.

Session 1: Seafood Smorgasbord

1. Have students work in small groups to brainstorm ocean animals that people eat.

2. Discuss and post their ideas.

3. Prepare students for Seafood Smorgasbord. Explain that as they sample, they'll try to guess and draw how each organism looked while alive.

4. Set up smorgasbord using the prepared trays.

5. Divide students into groups and have each start at a different table. Have them taste the unlabeled sample at their table, research it in the book (if available), and draw a picture of what they think the animal looked like.

6. Rotate the groups every five minutes (remind students to take their drawings with them) until all students have sampled from each table and illustrated their guesses.

7. After the smorgasbord, go back to each item and let students share their guesses. If no one guessed correctly, show a picture of the animal and tell them its name.

Introducing Thought Swap

1. Explain that in Thought Swap students will take turns talking with different classmates—cooperating, following directions, and talking quietly with each partner.

2. Have students stand in two parallel lines, each person facing a partner.

3. Say you'll be posing a question or idea for them to discuss with their partners for about a minute before "swapping" roles.

Thought Swap Begins

1. Begin Thought Swap with this two-part question:

- **How often do you normally eat seafood? What's your favorite kind?**

2. Help partners get started. When you call time, have a few students report something their partners told them.

3. Have one line move one position to the left, so everyone faces a new partner (person at end of line walks around to the beginning). Ask the next question:

- **What kinds of fish have you caught or seen someone else catch?**

4. After calling "time," have a few students describe their partners' responses.

5. Continue sequence with each question or idea below:

- **Describe how you think the seafood you just tasted may have been caught.**
- **Think about the activity Apples and Oceans. Where do you think most of the fisheries of the world are located? Why?**
- **What can people do to make sure enough fish will be around for future generations to eat?**

6. After the last question, spend extra time discussing it. Record students' responses on chart paper, to be revisited later.

World Fisheries

1. Distribute and have students read (in class or as homework) the Dolphins and Your Tuna-Fish Sandwich handout.

2. Discuss the questions on page 99.

3. Describe the three major problems facing fisheries: ***overfishing, habitat destruction,*** and ***bycatch***. Ask students to share proposed solutions.

4. Introduce ***maximum sustainable yield***, or ***MSY:*** total amount of fish that can be caught and still leave enough to reproduce, so the population doesn't continually get smaller.

5. Paraphrase Overview of the World's Ocean Fisheries (optional) or just distribute to each small group in Session 2.

Session 2: Fishery Experts

Fishery Information Cards

1. Divide class into five groups. Each group will become the "expert" on a different fishery and present its findings at a World Fisheries Conference.

2. Give each group a set of six Fishery Information Cards for one species, and six sheets of lined paper. Distribute an information card to every student in every group; point out different icon or drawing on each.

3. Have students read their cards silently and write down two or three important or interesting points.

4. Have each student summarize her card for her small group. Circulate to check for understanding and answer questions.

Fishery Assignments

1. After small-group discussions, distribute one packet of Fishery Assignments to each group. Each student picks out the assignment **with the same icon and name as his Fishery Information Card.**

2. Give each group one Overview of the World's Ocean Fisheries (copied on colored paper) as table reference while students work, and distribute grid paper for those with graphing assignments.

3. Direct students to poster supplies. Remind them some assignments include a form of graph as well as other representation.

4. Students may need to complete posters at home. In the next session they'll present them at the World Fisheries Conference.

Session 3: The World Fisheries Conference—*Poster Presentations*

1. After all posters are completed, have each small group discuss its fishery using pre-conference questions on page 99. One member of each group should record ideas.

2. Have each group make a presentation to the class as the "panel of experts" on its fishery, each poster in turn.

3. After all groups have presented, discuss what all the fisheries have in common.

4. Collect and compile all posters into the *World Fisheries Conference Proceedings.*

Session 4: The World Fisheries Conference—

"Big Picture" Recommendations

1. Prepare class to make recommendations at the World Fisheries Conference, working in the same fishery groups as before.

2. Distribute to each group one sheet of lined paper on which recorder can write the group's recommendations. Have them begin.

3. List all group ideas on chart paper, putting a star by overlapping recommendations. Each group can add to its own list as class list grows. Several possible recommendations appear on page 100; mention these if you wish.

4. Bind each group's list in the *World Fisheries Conference Proceedings.*

Think, Pair, Share: My Values and Behavior

1. Prepare for Think, Pair, Share, to review what students have learned and concluded about fisheries.

2. Distribute lined paper to each student. Have students **"think"** about and list all five fishery species from the World Fisheries Conference. Will they eat that seafood in the future? Why or why not?

3. For **"Pair"** and **"Share,"** have students discuss and compare their answers with a partner.

4. Tally the results by species into class bar graph. Do students believe their personal choices make a difference?

5. Distribute blank paper and colored markers to students and ask them to design individual mascots or slogans that support their viewpoints.

6. Display completed posters. Have half the students stand by their posters while the other half mingles and asks questions, then switch.

7. One at a time, hold up Key Concepts for this activity. Have students read aloud. Briefly discuss these statements' relevance to what they've learned about fisheries.

8. Revisit Key Concept from Activity 1: Apples and Oceans, and discuss this concept in light of What's the Catch?

There's plenty of fish in the ocean
And oodles of shrimp in the sea,
But due to the way we've been fishing
There's a ***whale*** of an e-mer-gen-cy.

Be it fish sticks, or tuna, or pollock
Halibut, haddock, or squids
Seafood's a main source of protein
For millions of fast-growing kids!

As more and more people inhabit
This planet of ocean and land
They'll need to keep themselves eating
So more and more food they'll demand.

But unless we take care of the ocean
And the teeming of life in its flow,
We won't have the food for the future
And down to the bottom we'll go.

We've taken more fish than we should have
Their numbers have drastically shrinked
"Sustainable yield" is exceeded
Some species may soon be extinct!

We've invented all kinds of devices
To rake up the vast ocean floor
The more we wreak habitat havoc
The bigger the problems in store.

We need to keep track of our fishing
Take only so much and no more
The fish can maintain populations
If we don't bring too many to shore.

While "bycatch," the word, may sound harmless
It covers up cruelty and waste
'Cause dolphins are captured and murdered
As tuna nets grab them in haste.

Some folks shrug their shoulders in sorrow
And others may just shake their heads
But a new spirit rose in young people
In a movement for CHANGE that they led.

And thanks to campaigns they have started
Now "Dolphin-Safe" tuna's on shelves
We need more of this kind of action—
More friendly environment elves!

Let's face it, the Earth's mostly ocean
All life sprang from its chem-is-try
If we don't save our ocean resources
It's down with the ship, you and me!

And it's true that there's only one ocean
Under sky that's a beautiful blue
Together we need to protect it
For without the sea what would we do?

The ocean is always in motion
And fish swim alone or in schools
If we don't keep the sea full and healthy
It won't be the *fish* who are fools!

There's plenty of fish in the ocean
And oodles of shrimp in the sea
But because of the way we've been fishing
There's a ***whale*** of an e-mer-gen-cy.

So learn about fishery wisdom
Do the math and then do the right thing
Make sure overfishing is ended
So tomorrow new harvests will bring!